Good-Bye, Boise... Hello, Alaska

By Cora Holmes

The true story of a family's move
to a remote island ranch.

Publisher: Roy J. Reiman
Editor: Mike Beno
Assistant Editors: John Schroeder,
Kristine Krueger, Henry de Fiebre
Art Director: Gail Engeldahl
Art Assistant: Julie Wagner
Production Assistant: Judy Pope
Photography: Julie Habel, Milt Holmes
Photo Coordination: Trudi Bellin

© 1994 Reiman Publications, L.P.
5400 S. 60th St., Greendale WI 53129

Country Books
International Standard Book Number: 0-89821-128-X
Library of Congress Catalog Card Number: 94-67825
All rights reserved
Printed in U.S.A.

For additional copies of this book or information on other books, write: Country Books, P.O. Box 990, Greendale WI 53129. **Credit card orders call toll-free 1-800/558-1013.**

Contents

Chapter   **Page**

Introduction 8
1 Beginner's Luck....................................11
2 Riding to Lamb Camp...........................19
3 A Trip Across the Harbor 31
4 Helga's Visit...41
5 Winter Water..53
6 The Fly Died..65
7 A Winter Baby......................................81
8 A Barn for All Seasons..........................91
9 Milton's Magic....................................107
10 Footsteps...117
11 A Hard One to Lose.............................131
12 A Long Way from My Heart....................147
13 One Good Splash Deserves Another........159
14 The Mall at Chernofski..........................175
15 A Mother First....................................185
16 The Mighty Angler...............................199
17 "We Can't Make Him Any Deader"...........209
18 Eternal Vigilance.................................221
19 The Good Samaritan.............................231
20 "Who Am I?" 243
21 In the Blink of an Eye...........................253
Epilogue...273

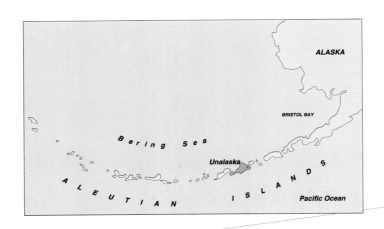

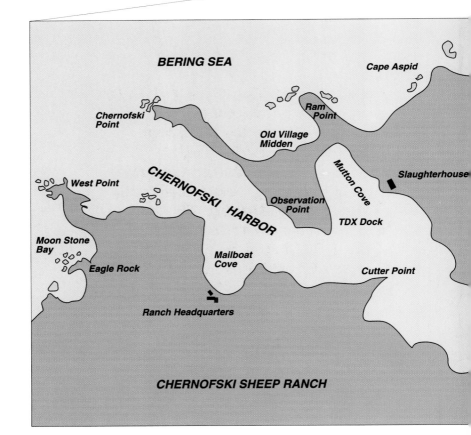

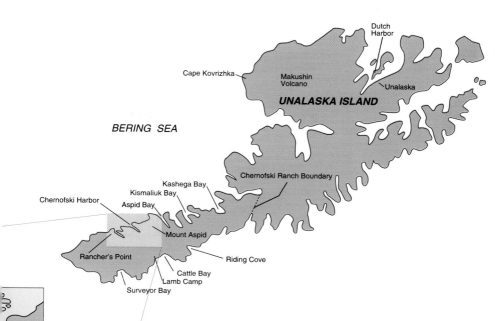

Dutch Harbor

Cape Kovrizhka

Makushin Volcano

Unalaska

UNALASKA ISLAND

BERING SEA

Chernofski Ranch Boundary

Kashega Bay

Kismaliuk Bay

Chernofski Harbor

Aspid Bay

Mount Aspid

Rancher's Point

Riding Cove

Cattle Bay

Lamb Camp

Surveyor Bay

UNALASKA ISLAND, Milt and Cora's home, is in the Aleutian Island chain, the "tail" that extends into the Bering Sea from mainland Alaska. The ranch was established around 1918 and encompasses 152,000 leased and deeded acres. It is 85 miles from the nearest town, Unalaska. Milt came to the ranch in 1948 to work as a foreman. He bought it in 1964.

Introduction

When an overprotective mother and her two sons from a town in Idaho find themselves transplanted to a 200-square-mile sheep ranch in Alaska's remote Aleutian Islands, they discover the challenges and choices of total isolation can generate dangerous, exciting and sometimes hilarious consequences.

With a wise new husband and father who lets them make their own mistakes, they learn to depend on each other and themselves, make decisions that could cost them their lives and discover the satisfaction that comes from hard work.

During the first year, every day brings a new adventure—riding horses, rounding up cattle, shearing sheep, trapping fox, beachcombing, operating an open boat in Alaska's stormy Bering Sea, going to school at the kitchen table, living without television, shopping out of catalogs and learning to love the sound of silence.

The lessons they learned were difficult. Sometimes they cried and screamed and sulked. But more often, they laughed and sang and joked. Most important, they did it as a family. Together through peace and war and truce, they were brought closer to understanding the real worth of the other.

On the surface, this is a collection of entertaining stories. But out of them emerges the daily tenor of a very different lifestyle. And between the lines runs a theme of family values, their importance and their strength.

Amid the stark beauty of the northern tundra, between cobalt-blue water and stone-gray sky, this new family grows close and forges bonds that last forever—with each other and with the island.

This is their story.

Acknowledgments

FIRST, I WOULD LIKE to thank Roy Reiman and the *Country* magazine subscribers who wrote and encouraged me to write this book. Your confidence and support helped more than I can describe in words.

Next, I want to thank my sons, Chuck and Randall, for allowing me to splash their youthful triumphs and tragedies across these pages.

And always, my heartfelt gratitude to Milt, who made all these adventures possible by marrying me and bringing us to this wonderful place, not to mention all the wood he chopped so I could burn the midnight oil.

Dedication

I'D LIKE TO DEDICATE this book to all the young men who served in the Aleutian Islands during World War II. Your legacy of freedom will always be a part of Chernofski. Wherever you are now, I hope you enjoy reading about, and perhaps recognizing, locations that haven't changed in the last half century.

Cora Holmes
Chernofski Ranch, Unalaska Island

MOVING DAY. Randall (top left) was 10 and Chuck 14 when they packed up to leave the motel in Seattle in October 1979 for their flight to Alaska. After settling in, airplane rides were rare events, as horses provided the main transportation at Chernofski Sheep Ranch. Above, Randall, Cora and Chuck are about 5 miles from headquarters on their way to round up some steers on New Year's Day 1982.

Chapter One

BEGINNER'S LUCK

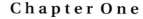

Chuck's horse, "Grey", stumbled to his front knees on the steep, narrow trail in front of me.

"Jump off!" I shouted, startling my own mount so that he skittered sideways on the loose rock. As I held onto my saddle horn with both hands, Chuck pulled Grey's head up firmly and got him back on his feet.

"What happened?" I asked as Chuck made a slow turn on the hillside and came abreast above me, his good-natured face already ruddy-cheeked from the cold.

"I don't know, Mom. Grey does that all the time." He grinned and patted his horse's neck. "I guess he feels the sudden need for a word of prayer and down he goes, to his knees."

"Well, be careful. You scared me to death."

"Sorry," he said, craning his neck. "I think I see them coming. We'd better spread out." With a *click-click* to Grey, he cantered away.

"Walk," I called after him.

What a difference, I thought, as I watched Chuck guide his horse carefully across the rocky frozen terrain studded with tussocks of brown grass, lichen clumps and moss. Just 2 years before, this confident 16-year-old was a skinny white-faced boy who had gotten airsick just before the floatplane landed on the bay in front of Chernofski Sheep Ranch.

My son Randall, meanwhile, was another matter entirely! He'd bounded off the plane with his older brother's new fishing pole and broken it in four pieces before Chuck's feet ever touched the ground.

A distant "Whoop!" alerted me that my 12-year-old was approaching. Bull-headed, determined, incorrigible and precious, he was Milton's shadow from the moment they met. At 10, Randall was ready for a hero and Milton Holmes was sure enough hero material. Fifty-eight years old, slim, wiry, ginger-haired, with a nose too big for his thin face, Milton walked into our lives with a slight limp and the kindest green eyes I'd ever seen. His kindness drew us like a magnet—even his dogs loved him.

While I waited for cattle and horsemen to appear on the skyline of Anderson Butte, I wondered again at the fate that prompted me to answer a "help wanted" ad in a Boise, Idaho newspaper in August of 1979.

At 33, I was ready for a change from working the night shift in Boise's neonatal intensive care unit. Chronically tired, always irritable from a high-pressure job, I was a miserable mother and I knew it. The position of housekeeper and secretary for a remote ranch on an island I'd never heard of sounded perfect. The minute I stepped off the plane, I knew I was home. After 16 months, Milton and I were married.

A stiff breeze lifted the hair on my cheek. I thought about where we were on the earth's surface. It still seemed like a dream that I might waken from and be in a traffic jam somewhere, with horns blaring and tires screeching. Chernofski Sheep Ranch—a tiny dot on the map, closer to Siberia than to the continental United States, enclosed by the Bering Sea to the north and the Pacific Ocean to the south, buffeted by gales, hidden by fog, jarred by earthquakes—was the most peaceful spot I'd ever found.

The remote camp on this mountainous island was little changed by the age of technology, and moved at the gentler pace of 100 years ago...

But, it didn't pay to dream while riding. My gelding, "Stormy", took advantage of my inattention to grab a bite of dry grass and I almost slid off over his head. Horseback riding was not a required course in nurse's training.

I was a nervous cowgirl, but if I wanted to watch my children grow up, share my husband's life, be a real partner and actually help, I had to not only get on a horse but I had to do things with it—like charge after fleeing animals, jump over creeks, plunge down hillsides, and ford creeks with big chunks of ice in them. Worse, I had to watch my children do the same. Unlike me, they had no concept of their own mortality.

"Heads up, Mom! Here they come!" Chuck's warning floated to me on the stiff breeze. Not a moment too soon, either, as silhouettes of animals and riders appeared on the crest of Anderson Butte above the trail. The cattle broke away and streamed toward us, Milton behind them, with Randall stuck like a burr close to his side.

I counted 30 head of mixed Herefords plus two big bulls, the rest steers and cows with calves. We needed five more steers to fill a fall order. The week before, we'd butchered six and sent them to our closest neighbors in the village of Unalaska, 85 miles away by boat. The beef was for crews on fishing boats. Together with the village, they comprised our market.

When the cattle saw Chuck and me blocking their way, they veered left and hit the trail right where Milt told us they would. With a sigh of relief, Chuck and I fell in behind them.

Our relief was short-lived. Milt and Randall picked their way down the steep hillside and caught up with us, Randall beaming, hatless, with one glove missing and a smear of strawberry jam across his chin. Perpetually hungry, he'd already been at his lunch.

"I helped, Mom! I really did! 'George' and I headed them off, didn't we, Milt?" Randall patted the raw-boned sorrel's neck and looked at Milt for confirmation. "And we found some sheep, too, over by the drift fence. Tell her, Milt."

I handed Randall a fresh glove from my saddlebag and made a swipe at his face with my mitten, but missed because George, seeing an opportunity, attempted to bite me on the leg. I couldn't understand why Randall liked that big ugly horse. He was clumsy and evil-tempered and had his ears laid back so often we called him a snake.

"I like him," Randall defended his choice. "He bites the steers on the rump and makes them go."

Milt reined "Fuzzy" to a stop beside me. "Can you get these cattle to the slaughterhouse? I want to take Randy and get that bunch of sheep."

My head jerked up in astonishment. But he wasn't looking at me; he was talking to Chuck. I saw Chuck's eyes widen, his throat bob on a convulsive swallow. Cautious and quiet, I wondered how he would react. His gaze followed the disappearing herd. "I guess I can try," he answered, his voice uncertain.

"You can do it," Milt encouraged. "Take Mama and don't let them break away into the swamp." He turned his horse, gave

RANCH HEADQUARTERS. Snug in Mailboat Cove, Chernofski Ranch headquarters is protected from the storms of the Bering Sea most of the time. The ranch is a welcome sight for the Holmes family after a day or more of work on the range or the water.

me a quick grin, motioned to Randy and off they galloped, back the way they had come.

"Don't expect any miracles," I hollered after them. They were already too far away to hear.

"Come on, Mom," Chuck urged. "Hurry." He rode downward, off the steep hillside.

I sucked in a scream as I watched Grey stumble on the slope, skid and put his front feet together. He slid the remaining few yards to the bottom with his rump on the ground. No way would Stormy and I do that.

I threw Stormy's reins over his head, kicked my feet out of the stirrups and jumped off. I came down the hill stiff legged, with Stormy snorting and clattering rocks just above my left ear.

By the time I reached the bottom, Chuck was a small blur in the distance, but he'd caught up with the cattle. He meant to get them in, I thought, and he didn't expect any help from me.

A burst of pride overrode my own fear. I had to get closer. Tightening my lips, and my cinch, I clambered back on my horse and gave him his head. Stormy didn't need any urging. He hated being away from the other horses.

As we rocketed down Chernofski Creek, I thought, for about the millionth time, how much more fun this would be if I were home reading about it in a book. "Armchair cowgirl," I taunted myself. "Remember all the years you and the boys dreamed of living this way? Now it's happening—enjoy it."

"I would, if only it weren't so high and so fast and so bumpy," answered a small voice in my mind. With a feeble show of bravado, I dug my heels into Stormy's sides. He gave a final burst of speed and, as Chuck and the cattle reached the head of the bay, we caught up.

Chernofski Harbor, shaped like a diagonal "J", snakes inland about 4 miles. The right side of the "J" consists of tidal flats and above them is the big swamp. Without a pause, the cattle rushed into it. Mud flying, tongues hanging out, sinking up to their knees, they didn't even slow down.

"Chase them!" Chuck screamed. "If they stop they'll bog down!" Shouting and whistling, he forced Grey into the marsh. Stormy wouldn't go—he refused to budge. I wasn't strong enough or brave enough to bully him into a place he knew wasn't safe. I bailed off again and dragged him around the edge, shouting and screaming at the cattle, at Chuck, at the horse.

"Move along, you stupid cows! Get out of there, Chuck, you'll drown! Come on, Stormy, jump!"

Chuck raced back and forth in a frenzy, slashing rumps with his reins, hollering, whistling, swearing. When the last bull clambered to safety, Chuck's face was bright red, covered with sweat, triumphant.

Back on the trail, we were within 2 miles of the slaughterhouse corral. We had a choice—over the hill on the old military road or around the beach. The hill was shorter, easier on the horses, but held a dozen escape routes. The beach was rocky, littered with driftwood and spikes, hard on horses hooves and slow going. But the cattle couldn't get away from us. We chose the beach. As soon as we got the cattle pushed across Pump Creek and onto the beach trail, they bolted out into the bay and stopped in 2 feet of water.

"Let's rest," I panted.

"Okay, but only for a minute," Chuck agreed.

We stayed on our horses. Milt had warned us enough times about the unpredictability of

HEAD 'EM UP! *Randall never had to worry much about getting cattle to move when he was riding George. If the cows didn't get along, George gave them a nip in the shanks to help them. Randall was only 12 here, but already a cowboy.*

range cattle. As we rested, I looked across the bay at the twinkling lights of seven floating crab processors anchored in the deep water. The Dutch Harbor king crab season was open and these huge boats were waiting for fishermen to bring them their catch. Alaskan king crab, a new multi-million dollar industry, was right here on our doorstep.

We benefited from the boat traffic in an increased market for beef, more frequent transportation, occasional mail and a sideline business storing crab gear for the fishing boats. Although our main cash crop was wool, we laughed about all our other small business ventures and called ourselves "Podunk Enterprises Inc."

"Time's up," Chuck said, and whistled at the cattle. But they were hot and tired and wouldn't move. We hollered and threw rocks until our arms were numb. When they did move it was at a walk, and most of the way around the beach they kept getting back into the water.

By the time the slaughterhouse was in sight, both Chuck and I were tired and hot. The closer we got, the more cross Chuck got. Soon we'd have to force the cattle up the beach and out into the open before we could get them through the gate. If we lost them there, they could run clear out on Chernofski Point, 2 miles away, and we'd never get them.

Chuck left me behind and loped away to open the gate. I pushed the cattle ever so gently up onto the road. As we waited for the lead cow to take them through the gate, Chuck was a model of patience.

He tossed small pebbles that landed behind the animals, then waited for them to move an infinitesimal inch closer. Not until the last shaggy red coat was behind bars did he let out a war whoop and slam the gate shut.

I slumped in gratitude over Stormy's neck while Chuck's whole personality changed. He leaped into his saddle.

"I suppose you're too tired to trot home," he said with a grin and fished in his saddlebag for the lunch both of us had been too busy to eat.

"I can do anything you can do," I lied. Now that the cattle crisis was over, my mind leaped to the next ordeal. The ranch headquarters was at least 5 miles away and it was nearly dark. "You lead the way."

When he saw I wasn't going to fall off at a trot, Chuck kicked Grey into a gallop. I kept up by making bargains with myself. Please, God, I'll never swear again and I promise I'll sweep under all the beds. No matter what happened, I wanted to be at the narrows before darkness fell. If the tide was in, we'd have to swim our horses, and the thought terrified me.

We made it, but just barely. It was too dark for me to see the slow-moving ice chunks that kept bumping into Stormy's legs, but I knew they were there by the way he danced sideways.

The wind had blown ice into a thick wall against the opposite bank, and both horses reared and shied away from it. Chuck coaxed Grey up onto the slabs and grabbed my reins and pulled me behind him. The rest seemed easy, straight up a narrow trail full of gullies with soft bottoms and big boulders.

We usually got off and walked, but not this time. We had to trust the horses. Island natives, they found the right footing and fetched us up against the high gate of Cutter Point pasture. From there to the barn was a mile, and knowing a manger of oats waited for them, they shot away like 3-year-olds out of the starting gate.

Nothing was quite so rewarding as coming off that last hill and seeing the lights of ranch headquarters shining from all the windows. It meant Milton and Randall had gotten home safely. They would have a fire started. The house would be warm.

We reined our horses to a walk and splashed through the last creek with the lights getting closer and closer. The ranch looked like a miniature village. A warm, welcoming home.

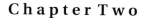

Chapter Two

RIDING TO LAMB CAMP

I stopped in mid-reach for my saddle when I heard Chuck and Randall whispering from the far end of the tack room. In the dim light of this winter morning, they hadn't seen me and their earnest discussion covered my footsteps.

"Lamb Camp is an easy trail, so you watch out for Mom today, okay?" Chuck's deeper voice carried to the entry where I eavesdropped shamelessly. "It's my turn to ride with Milt."

Randall's reluctant consent surprised me almost as much as discovering the boys thought I needed looking after. Here I'd been congratulating myself for not falling off my horse or getting lost in the fog, and all the while they were taking turns baby-sitting me. I had to smile—they knew me pretty well.

I lifted my saddle off its tree and the noise stopped their conversation. They scrambled for their gear and bolted out of the barn.

While I didn't agree with Chuck that Lamb Camp was an easy trail, it was one of my favorite rides. I was kind of glad Milt and Randall hadn't gotten the sheep out of there yesterday, and that all four of us were going back today.

Six miles away from ranch headquarters, southeast across the island on the Pacific Ocean, with its windswept dunes of black volcanic sand and Stonehenge monoliths of gray basalt, it drew me with its otherworldly atmosphere.

Violence and upheaval were stamped on its landscape as clearly as the stories Milt told about it were etched in my mind. In the early days of the ranch, it had been used as winter pasture for weaned lambs, complete with fences and a line cabin. When a tidal wave destroyed the camp in 1956, washing away the tundra and scattering driftwood half a mile inland, it uncovered an ancient Aleut village site where human skeletons still sometimes surface through the sand.

It often seemed a sad, lonely place, with gulls crying from off-shore rocks and breakers rolling endlessly over the wind-blown sand. Still, I loved the harsh beauty and stark atmosphere. Perhaps because Lamb Camp was the first actual ride Milt ever took me on; the beginning of our romance.

I remembered the joy of that day as the four of us walked together with our backs to the wind, trying to keep the sand out of our peanut butter and jelly sandwiches. Even knowing the only fatality in ranch history had occurred there didn't diminish its pull.

I tightened my cinch and patted Stormy's neck. None of us were careless enough to get lost. Besides, I never let either of the boys out of my sight unless they were with Milt. Even then, we were always within hollering distance. And if the worst happened and one of them did get separated from us, the horses knew the way home. All they had to do was give him his head and

RUGGED BEAUTY. Although rides to Lamb Camp and other spots on the island are often arduous, they are made easier by enjoying the breathtaking beauty of Unalaska.

he would be back at the barn with his nose in the feeder before the rest of us knew he was gone.

Out of the corner of my eye I saw Milt unobtrusively check the cinches on Chuck's and Randall's horses. Satisfied, he walked over to where I had Stormy tied to the wagon tongue. With a slight smile, he undid my knot and pulled Stormy up another notch. He didn't say anything; after 2 years of marriage, I was used to his quietness. He gave me a hand up and squeezed my leg through the thick gum boot I wore. That said everything.

"Time to go, boys," Milt called as he swung his leg over Fuzzy with the slightest grimace. "Those sheep we saw yesterday are wild, so save your horses. We'll have to ride hard to get them in." He whistled for the dogs and nudged Fuzzy into a fast walk.

As we rode away from headquarters, I watched Milt shift around in his saddle trying to ease the strain on his bad leg. Two years before I met him, he had his right hip replaced because constant riding had destroyed the joint. The surgery was still comparatively new and the discomfort from the prosthesis was sometimes as bad as the pain from the joint.

I hoped we'd have an easy day. At least the weather was cooperating. No snow on the ground yet, which was nice for early January, and the temperature was above freezing. A few raindrops spattered us, but they were occasional and the sun was trying to break through the overcast.

The first part of our ride was a gradual 2-mile climb, taking us from sea level up toward Foggy Butte, which, at 843 feet, towered above the ranch and formed our southern skyline.

Since we live in a world without libraries, churches or neighbors, places take on special significance. Chuck loves to watch the waves crash against the cliffs on West Point. Randall loves Konet's Head, where he can watch the wild horses. Milt loves the ancient village sites with their low rock walls and bleached whale bones.

And I love Foggy Butte. It is my cathedral. Rising steeply beside a deep quiet lake, it juts toward the sky. Whenever there is a patch of blue in our foggy skies, it usually rests above Foggy Butte.

As we maneuvered our horses along the narrow trail above the lake, I lifted my eyes to the skyline and picked out the rough stone outcroppings I leaned against on Sunday afternoons. No church in the world inspired me as did this beautiful gift of nature. I got a lump in my throat just looking at it.

With Foggy Butte behind us, we descended into the wide Kranich Valley, named after a ranch manager from back in the '30s. During summer, the grass is thick and green here, with wildflowers everywhere. In winter, the grass is brown and the flowers are gone, but the protein is still there. The sheep we sought would be fat on the winter grass. When we reached flat ground, the horses speeded up. "You boys slow down," I called as I sawed on Stormy's reins. They both circled back and fell into step beside us.

"You know, Mom," Chuck complained, "if we robbed banks, we'd never get away because we wouldn't even know how to gallop our horses."

"Yeah, Mom." Randall agreed, pointing his finger into the air and firing an imaginary shot. "Bang! Bang! We're the Chicken Gang." He urged George ahead. "Come on, Chuck. Let's git!"

They trotted just ahead of us with their voices floating back over their shoulders. George stumbled and Randall lurched over the saddle horn, righting himself in an instant, hauling George's head back up, never missing a beat in the silly song they were singing.

Milt's hand on my arm stopped me from scolding. "Let them be kids," he said. "I like to hear them laugh."

"I don't want them hurt," I protested.

"They're safer here than anywhere else," Milt answered. "No traffic, no germs, no drugs, just clean air and space."

I left my fears unvoiced. I, too, liked to hear my sons laugh together. More often they quarreled and bickered. With 4 years difference in their ages, and opposing personalities, they seemed to rub each other wrong most of the time. Chuck was quiet and dreamy while Randall was volatile and boisterous. Only their immediate unity when outside danger threatened convinced me they truly cared for each other.

We climbed out of Kranich Valley and the topography changed. The high ground on the ridge top was rocky, grassless and covered with scree. The wind always blew up here, even on the calmest summer days.

"Time to stretch our legs and give the horses a break," Milt said. He dismounted and we all followed him. The narrow path downward was steep and twisting. I took my time picking good footing and soon lagged behind the others.

When I heard a commotion on the trail ahead, somehow I knew Randall would be involved. Muffled shouts, sounds of

MILITARY MEMORIES. Back in the '40s, the military had a base on this isolated island. Some of the buildings are still used in the ranch operation, but many, like this Quonset hut, have succumbed to the harsh Aleutian weather.

struggle, a horse's frightened whinny. I started running.

The scene I came upon bore out my worst nightmares. George was upside down in a deep narrow creek. He squealed and grunted as his legs pawed the air and his whole body twisted. Where was Randall?

My eyes searched frantically. His thin wiry body crouched at George's head. Tears streamed down his face.

"That's right, Randy," Milt said. "Keep his nose out of the water."

As Milt talked, he uncoiled the rope he always carried on his saddle. "I'll tie this to George's tail," he told Chuck, who stared at the horse with funeral eyes. "When I tell you to go, lead Fuzzy back uphill toward Mom."

Chuck grabbed Fuzzy's bridle while Milt's hands knotted the rope. "Okay, go." Obedient to the slightest pull, Fuzzy followed Chuck up the hill. The rope tightened on George's tail. "All right," Milt said. "Pull on his halter rope, Randy."

Randall scrambled to his feet and put all his strength into a gigantic pull. "Get up, George, before you drown," he pleaded. The horse, whose frantic attempts had ceased, renewed his efforts. With Fuzzy pulling him onto his side, George got his legs under him again. After one mighty lurch, he was back on his feet.

"Get back!" Milt shouted. He jerked the rope off George's tail and the horse clawed his way up the steep bank, flinging water and mud in all directions.

"Good boy!" Randall yelled. "You can do it." He dropped the halter rope and dashed to one side. With a final lunge,

BERING TREASURE. The sea provides more than halibut and crab for the fishing boats that anchor offshore from Chernofski Harbor. On those infrequent days when they take time off, Cora, Milt and the boys beachcomb. Among the treasures they find are glass floats from Japanese fishing nets. Today most of these beautiful glass balls have been replaced by plastic, so those at the ranch have become real collector's items. It is also not unusual to find bottles afloat, some with notes in them dating back as far as 20 years.

George topped out on level ground, gave himself one colossal shake, lowered his head and began cropping dry grass as if nothing had happened.

"No harm done," Milt said as he ran his hands over George and checked Randall's saddle. "You might have squashed sandwiches, though." Randall paid no attention; he was too busy hugging George's neck and patting his nose.

Reaction set in; I started to tremble. "What happened, Randy? How did George get in the creek?"

Randall shrugged. "He just fell in, I guess."

"I saw the whole thing," Chuck interrupted. "It wasn't Randy's fault. George stepped in a hole or something."

Milt walked along the bank. "Yup, right here." He pointed to the broken-off bank. "You can't trust these land bridges. It's best to go around them." He grabbed Fuzzy's reins and led him to a spot farther upstream where all of us crossed easily.

"We can ride for a while." Milt mounted and led the way on down the steep trail. And that was that. No recriminations, no stern lectures.

While we ate our lunch on the sandy Pacific beach, I thanked Milt for being so nice. He looked down the beach where the boys were searching for glass floats, then back at me, in surprise.

"Why scold? Randy learned a good lesson about crossing creeks without me saying one word."

He was right. "How do you know so much about kids?" Most of the time I felt helpless and lost trying to raise the boys.

"I got in plenty of practice on my own two," he answered. "I made mistakes, too, but they always knew I loved them."

I didn't know his sons well. Already adults when I met them, our paths didn't cross often. His oldest, Stanley, was 30-something and the mayor of Unalaska village. Val, 4 years younger and a commercial fisherman, came to visit us at the ranch on rare occasions. Randall tried to pry stories of Val's childhood on the ranch out of him, but he was even quieter than his father, if that was possible.

The minute we finished our lunch, we set off after the boys. The half hour we allowed ourselves to beachcomb was a real high point of our long day. I found a glass sake container from

Japan with a flower painted on one side. Milt found a hardwood board with three stainless steel screws in it. He spent the rest of his time getting them out with his pocketknife.

The boys returned with one small netted glass float, the kind used by Japanese fishermen before plastic became common, and a wooden stamp with Japanese symbols on it. By the time we got our treasures packed in our saddlebags, it was time to leave.

Milt had spotted a small band of sheep with his binoculars high on the hillside to our left and we were going after them. Once away from the beach, he turned to the boys. "One of you ride up to the high trail and head anything you see toward the drift fence. We'll meet you there."

Without a moment's hesitation, Chuck kicked Grey's flanks and took off. Randy sighed but reined George in so he stayed beside me. Milt took the dogs up around the sheep he had seen earlier.

Soon the sheep were streaming toward us, so Randy and I spaced ourselves and rode to them. With three of us and the dogs, they weren't hard to control and we soon had them in a tight bunch headed the right way.

I kept glancing toward the high trail, expecting to see Chuck's sheep break over the edge. He must be having a hard time, I thought, as we rode along without seeing him. He doesn't have any dogs. I hope he doesn't ride all over and get so far behind us he can't catch up. By the time we reached the east drift fence and he hadn't appeared, I started getting concerned.

The sheep were still fresh and wanted to run; we couldn't wait. We loped along the fence for 2 miles before they finally pulled up and rested. Still no sign of Chuck. I turned Stormy and looked back down the fence. Nothing. "Shouldn't Chuck be catching up pretty soon, Mom?" Randy stopped beside me.

"Yes," I said. "I'm worried something has happened. Maybe his horse got away from him or he's hurt. Go see what Milt thinks."

Randall nosed his horse around the band to where Milt held back the lead. After a short conversation, he came back.

"Milt isn't worried. He thinks Chuck probably got a bunch spooked out on the high trail and they wouldn't break off. He thinks Chuck is ahead of us someplace."

I looked across the sheep at Milt. He must have seen the distress on my face. He pointed back the way we had come. "Go look for him if you want," he called. "But don't tarry."

I didn't need any urging. "Come on, Randy." We trotted our horses all the way back to the beach. We called and whistled and looked. We rode up to the high trail and searched every hillside. No sheep, no horse, no Chuck. The ground was rocky and impossible to track a horse in. Just when I thought we were following his trail, it would peter out.

Our voices were hoarse from calling. The sky had darkened to slate gray. If we didn't leave soon, it would be dark before we caught up with Milt. How could I leave without finding Chuck?

"We better go." Randy shivered under his coat. "He isn't here." I shivered, too, despite myself, but it wasn't from the

SOME SHEPHERD. Milt brings in a herd of sheep to the upper corrals. The animals provide food, clothing and income for the Holmeses. An experienced hand with the woollies, Milt could, with just his horse and three dogs, control a herd of 2,000 sheep!

cold. I was remembering the story of the only fatality that had ever happened on the ranch. That had been winter, too, during a bad storm.

Stacy Ugalde, a Basque shepherd, and a companion were caught afoot at Lamb Camp. Stacy, smart, experienced and well-acquainted with the terrain, tried to convince his companion they should stay together, that he, Stacy, knew the way back to headquarters. But the other man would not listen and plunged out into the storm on his own. Weeks later the man's dog returned to the ranch, but no trace of the man was ever found.

"Let's make one more quick pass," I said. We did, but it was fruitless. We turned our horses away from the Pacific and ran them along the drift fence. Darkness came so fast, soon we could hardly see the next ridge.

"What's that, Mom?" Randall pointed across the fence to the side of Anderson Butte we were passing. "Is it one of the dogs?"

A blurred black shape raced across the tundra. A fox. I sucked in my disappointment. "No," I told him. "It's a black fox." Somehow I had hoped it was Chuck. "Hurry up."

Randy loped beside me. "I think they call black foxes 'Silvers', Mom, and they're worth a bunch of money. Do you think this one will get close enough to the house for me to trap?"

"Maybe," I replied, listening with only a small part of my mind. Where was Chuck? Why hadn't he met us where he was supposed to? I couldn't believe he was already ahead of us. Without a dog and all by himself, he would never attempt to get a band of sheep in. It would be impossible. Yet I was sure he wasn't behind us.

We caught up with Milt just as he was going through the lower gate on Cutter Point pasture. "No sign of him, huh?"

"No, and we looked everywhere."

"Well, that means he must be in front of us somewhere. Come on."

We pushed our sheep as fast as they would go in the dark. When we reached the main trail, we got off our horses and walked. Milt knelt down and inspected the trail. "I can't be sure," he said, "but I think there are more tracks here than our little band would make."

I was so convinced that Chuck was lying in a gully somewhere underneath his horse that Milt's words didn't even sink in. When we came over the Cutter Point hill and looked down at ranch headquarters blazing with lights, I just stood there in a daze. But it wasn't until I saw Chuck grinning at us from the barn door that I actually believed he was safe.

"Hey, you slowpokes. I got home with 35 sheep at 4 o'-clock. I even have supper ready."

"Tell us what happened," Milt said as he helped me off my horse. Suddenly I felt very tired.

"When I got up on the high trail, this bunch of sheep took off

DIPPING TIME. As Cora watches, Chuck and Randall, on horseback in the distance, bring in a flock. The sheep have already been sheared and are being readied for a dipping. The sheep live a good life on the island, which has more moderate winters than interior Alaska.

right in front of me. They ran all the way home. I just stayed behind them until I got to the gate. Then I got them through the gate and they came right to the barn."

He tried to keep the pride out of his voice but it kept bubbling through. His smile was so big it threatened to split his face.

"Good job, Chuck." Milt said.

NATURAL FOOD. Kelp, washed in from the sea, provides feed for these horses. They aren't the only ones to benefit from the sea's bounty. When driven by the weather from their grasslands, cattle often come down to the shore for a meal of kelp.

"Well, I saw a silver fox on Anderson Butte," Randall said from George's stall. "And it's mine. I got my dibs on it."

"Not if I catch it in my trap," Chuck retorted.

"Yeah? Well you better not catch it, 'cause I saw it first."

"Yeah?"

"Yeah, I'll just take it if you do."

"We'll see about that."

"Oh, yeah?"

I sagged against the barn door and closed my eyes, letting the blissful sound of their bickering wash over me.

Home safe again...and back to normal.

Chapter Three

A TRIP ACROSS THE HARBOR

The crackle of radio static woke me to a pitch-black morning.

"You got this on, Milt?" The voice floated above our heads and I felt Milt stir beside me.

"This is Mike Lynch on the *Silver Clipper*. I'm coming in to get pots."

Our feet hit the icy floor on opposite sides of the bed at the same time. A match flared and I saw the plume of steam from Milt's breath curl up around the kerosene lamp. In the soft light, we pulled clothes on over our long johns. The clock said 6 a.m.; the sun wouldn't be up for another 4 hours.

While Milt felt his way into the radio room, I lit my lamp and hurried to the kitchen, thankful I'd taken time to rake down the coals and lay a fresh fire the night before. I dreaded getting up in the dark to a cold house.

All I had to do in the frigid pre-dawn was put a match to it. Some mornings I snuggled back under the covers for another half hour of cozy warmth before the fire was hot enough to cook breakfast, but not today. I already knew we'd be busy.

Milt wanted to butcher three of the steers Chuck and I had brought in. From the sound of his conversation on the radio, we'd add several hours of crab pot hauling to that job. Mike Lynch was one of our best customers and a good friend. He and his sons owned two crab boats, the *Silver Clipper* and the *Tanya Rose*, which fished the Bering Sea.

They lived in Seattle, and when they brought their boats out to the Aleutians, they'd always carry up our supplies—grain one year, groceries the next. One of the animals we'd butcher today was for them.

I heard Milt sign off before I stepped outside to get bacon and milk from the storeroom. Even before my feet slipped on the frozen walk, I knew the weather had changed. The wind was still, hardly enough to sway the steam from my breath, but the temperature had plummeted to 20°.

Back inside, the roar of the coal fire and the lamp's rosy glow enclosed us in a cocoon of comfort. I fixed breakfast before I woke the boys. They were tired from 2 days of hard riding and I had to shake them both awake. Like me, they didn't enjoy getting up in the dark and cold, but they were a little more vocal about it.

"It's the middle of the night," Chuck protested as I lit the lamp beside his bed and left it shining mercilessly in his face.

"Go away, Mom," Randall groaned, grabbing at the covers I pulled into a heap at the bottom of his bed with callous disregard for the goose bumps I caused. "I hate school." He hid his head under his pillow.

"No school today," I announced. "We have to butcher and haul crab pots for Mike."

Randall jumped up and dashed into the kitchen with his clothes. During breakfast, he kept up a steady stream of questions. "Did Mike bring the mail?" He swallowed two slices of bacon in one gulp and washed it down with a whole glass of milk.

"I don't know," I answered from the stove where I was cooking the last of the pancakes. "He didn't say."

"Can I drive the boat, Milt?"

"Not in the dark. Maybe on the way home," Milt answered, sandwiching fried eggs between sourdough pancakes and pouring maple syrup liberally over them.

"But you can drive the Bobcat from the slaughterhouse to the dock for me when we start hauling pots." He slid bacon onto the rim of his plate and handed the dish to Chuck, who sat quietly in the shadows of the lamp.

"Super!" Randall shouted, spooning up the last of his oatmeal and reaching for the pancake platter. "Can I get on Mike's boat and watch TV?"

"You'll have to ask your mom about that."

"We'll see," I said, pouring a second cup of coffee and taking it to the sink with an armful of dirty dishes.

The water sputtering from the tap never fails to amaze me.

A BOUNTIFUL BREAKFAST is in the making by Cora, on what at first looks like a wood-burning stove…but there's a coal scuttle on the floor next to this stove. With no trees on the island, what firewood there is comes floating onto shore in the form of driftwood. A sizable coal pile remains available across the harbor, left there by the military in the late '40s. The fuel needed to keep the ranch warm and the stove cooking comes from that lode.

With only the help of gravity, coils circulating through the firebox and Milt's ingenuity, I always have hot water at my fingertips. Filling the sink with steaming soapy water, I called out, "Bring your dishes when you finish."

"Hey," Randall exclaimed as Chuck came into the lamplight with his dishes. "How come you're all dressed up?"

I glanced around, still mildly surprised to find myself looking up at this youth whose eyes had been level with mine a few short months ago.

It was true. Besides the clean shirt, Chuck had taken pains with his hair. Instead of the hit-or-miss treatment he usually gave the long, dark mass, it was neatly combed and secured with a bandanna around his forehead.

DOCKSIDE DELIVERY. When you live on an island, you get what you need by boat. Here the Tanya Rose, a fishing vessel, brings some supplies to the ranch. The vessels also provide customers, and income, as the captains buy beef and mutton to feed their crews.

Chuck glared at Randall. "I'm not dressed up!" he hissed. "Just because I put on a clean shirt doesn't mean I'm dressed up."

"Hah," Randall smirked. "You think there'll be girls from the processors walking on the beach today." He danced out of Chuck's reach, splashing his dishes into the sink as a reason to get closer to me. "Well, I'm going to tell them you're only 16."

"Who cares?" Chuck shrugged his shoulders. "They'll never believe you. You're just a kid." With a supercilious smile, he turned on his heel and followed Milt outside.

"It isn't fair," Randall complained after they'd gone. "All the girls from the boats think Chuck is 19. They're always hanging around him." Randall rubbed his face with a dejected sigh. "I wish I had a mustache—that's what makes Chuck look so old." He pushed out his upper lip. "When will I get one?"

I stared at him in astonishment. He really was serious. "You're only 12, son," I finally said. "Chuck didn't get one until he was 14, and there's no guarantee you'll be like him. You're adopted, remember. You don't have the same genes."

Randall had always known the circumstances of his birth, that he had been adopted when he was 3 days old. As far as I could tell, it had never bothered him. Now he just sighed again and nodded. "I can't wait 2 whole years," he said. "I need something now."

I groped for something to say that would cheer him up. "Those girls are much older than you are, Randall. All of them are over 18—most are in their 20s and 30s."

I wiped down the counters as he rinsed the last dishes and set them in the drainer. "Why do you want them to notice you?" I asked. Was he that lonesome out here?

"If Chuck can get girls, then I want girls, too," he said with a stubborn set to his chin. He dried his hands on his jeans and hurried after Milt and Chuck.

I added a bucket of coal to the fire, knowing it would never last until we came home, but hoped the water would still be warm in the tank and the house not entirely frigid. Then I pulled another sweatshirt over the one I already wore, poured the rest of the coffee into a thermos and blew out the lamp. Feeling my way out the door, I followed the others into the dark.

So that was it! I should have known. The competition between the boys was fierce, for my attention, for Milt's…and now for girls!

As I closed the horse pasture gate leading to the beach, I heard the marine ways engine start. I quickened my steps. The engine pulled our 21-foot, double-ended dory into and out of the water on miniature railroad tracks that extended into the bay about 25 feet, far enough to clear our 4-foot tide change. If I missed getting into the boat before it was lowered, I'd have to walk out to it on the slippery tracks.

I needn't have worried. When I got there, Milt and Chuck were prying with crowbars on the dolly wheels, which were frozen fast to the metal tracks.

"Help Randall," Chuck panted to me.

I boosted myself over the bow and rolled into the boat. Inside, Randall was rocking back and forth, so I added my weight to his. Between us we provided enough motion to jar the dolly beneath the boat and break the wheels free from the

AHOY, MATES! Milt stands next to the dory the family uses to get to other parts of the island when not riding horses. The plank is used to walk to the boat, which is lifted in and out of the water by a dolly. The ramp the boat is on is called the "marine ways".

frozen tracks. With a screeching lurch, the dolly started down the tracks toward the black still water. At the first motion, Milt dashed into the winch house to monitor the cable coming off the spool.

Randall and I toppled backward and collapsed onto the first wooden seat while Chuck vaulted over the bow. "Is the plug in?" he yelled.

Without waiting for an answer, I launched myself toward the stern, clambered over the second seat and motor well, my hands feverishly searching for the bolt that fit into the drainage hole in the bottom of the boat.

"I already put it in," Randall called out, just as the boat entered the water with a gentle slap and my fingers touched the familiar ring. I gave it another couple twists just to be sure.

When we felt the boat float off the dolly, Chuck signaled Milt with a flashlight and the winch motor stopped. Then, while Chuck and Randall untied the mooring lines, I held the flashlight on the tracks so Milt could see. With a 10-foot aluminum pike held like a tightrope walker's pole, he balanced on the slippery frozen track and inched toward the boat.

For the last 15 feet, the track had treacherous footing, covered with water and slick with dead seaweed. I kept the beam in front of Milt, biting the inside of my lip, releasing a pent-up breath when he stepped out of the freezing water into the boat.

As soon as he had the 25-horsepower outboard started, the boys let go of the mooring lines and we moved out into the smooth black water. By this time, my eyes were accustomed to the dark and I could see the outline of the headland rearing up ahead of us.

Without any wind to ruffle the bay's flat surface, the boat plowed through the water like a knife, leaving a wide V-wake behind. The boys sat quietly on the seat in front of me, the outboard noise discouraging conversation.

I looked at their straight backs and wondered again if I'd made the right decision. Was I cheating them out of a normal childhood by isolating them from friends and classmates? And what about their education? Was it fair to take them out of public school, where they had the benefit of up-to-date teachers, facilities and resources to stimulate their minds?

I thought back to the summer 2 years before when Chuck had graduated from eighth grade. I remembered the police car pulling up outside our house with him and a friend inside, their fishing poles beside them on the seat. The guilty yet defiant look on his face when the officer explained to me he had caught them throwing rocks at mailboxes hadn't disturbed me as much as his offhand rationalization, "Everybody does it".

Randall, meanwhile, was displaying disruptive behavior in the classroom and fighting on the playground. By the fourth grade, he still could not read.

Shoved into special classes, tagged as "different" by his classmates, defeated scholastically before his 10th birthday, he'd retreated into a fantasy world of television characters and video games.

As our boat rounded the headland and started through the channel into the inner harbor, I watched the boys' faces in the lights from the big crab processors we passed. Randall grabbed

Chuck's arm and pointed. His eyes shone with excitement.

No, I told myself again. I'd made the right decision and...

Thud! The boat suddenly shuddered and swerved in the darkness.

"Ice!" Randall shouted, peering over the side. The boat shook again and I heard the sound from the engine change. "Look!" he yelled. "It's everywhere!"

We were surrounded by great jagged chunks of slow-moving ice. They banged into us, shaking the boat, looking huge and menacing in the reflection from our running lights. I turned panicky eyes to Milt. His calm face didn't reassure me.

"What's happening?" I yelled.

"Don't worry!" he shouted back. "It's freshwater ice from the head of the bay." He stood up and swung the boat into an opening.

"Navigate for me, Chuck!"

The boat tipped crazily as Chuck lurched to the bow just as another solid chunk nudged us. Randall clambered over the seat beside me. The boat slowed to a standstill.

"This is neat!" he exclaimed. "We're iced in."

"Sit down!" I screamed. It took all my willpower to stay calm. My heart was bouncing inside my chest like a handball. I clung to the wooden plank seat. The sound of ice grinding against the boat's hull terrified me. When Milt's hand tapped my shoulder, I gasped.

"Don't worry." He spoke close to my ear. "I want you to rock." He made swaying motions with his body. "It helps." Randall's eyes lit up still more.

"All right!" he shouted. "Come on, Mom!" He lunged against the side of the boat making it tip sluggishly. "Rock!"

I nudged my body gingerly against the railing, sucking in my breath as the boat dipped, bringing the black water closer to my face.

"More!" Randall shrieked. "Like this." He slid across the seat and into me, pushing both of us against the side of the boat. The black water rushed toward us, so close I could count the individual ice crystals. Would they find our bodies? I wondered. When Chuck felt the motion of the boat, he added his own weight to our rocking. Since he couldn't shout loud enough for Milt to hear over the motor, he pointed the direction he wanted him to take; his arms flailed like a windmill.

The boat zigzagged as it rocked, and still the chunks rammed into us, crunched along our sides, swirling into our wake, hitting the propeller shaft with solid, sickening *thunks*. Behind me, Milt tightened his grip on the outboard's tiller.

Darkness enveloped us again as we crept away from the anchored crab boat's lights. I felt small and alone and far from shore. The *Titanic* came to mind.

Then, as suddenly as it had appeared, it was gone. A channel widened in front of us and we reached open water. The boat picked up speed. Behind, I looked at the floating mass of ice—the twinkling boat lights reflected off it in a million sparkling glitters. I wish I could say I appreciated the adventure and beauty of an incident most people never get to experience, but all I could think about was going back through that ice to get home.

The boys loved it. As soon as the chunks thinned out, Randall stopped rocking and scrambled to the bow where he and Chuck searched the water for ice floes.

"Look at the size of that one!" Chuck yelled.

Randall aimed his imaginary gun. "I wish we had polar bears."

In the still, cold dawn, I envied their excitement, untinged by any concept of mortality, admonishing myself for my own fears. Just then Milt squeezed my shoulder. "All clear," his husky voice whispered behind my ear.

Oh yes, I had made the right decision.

HEADED HOME. Nestled in the brown hills, the ranch headquarters is home sweet home. In the foreground, the tracked vehicle is towing a trailer...it's that or horses, because nothing with wheels works very well on the soft tundra.

Chapter Four

HELGA'S VISIT

We docked near the slaughterhouse safely, but as Chuck and I unloaded the boat in the semi-darkness, we had a mishap...I hit him right in the face with one of his tennis shoes.

Since I didn't want the shoe to land in the water, I had thrown it hard, then heard the solid *whack* as it connected with his cheek.

"Sorry," I mumbled, but couldn't muffle a chuckle when I saw the look of disbelief as he pulled off his glove and rubbed his jaw. After a moment, he grinned, too. "Good shot, Mom."

Soon, we were both laughing like idiots...after our ordeal in the ice, I needed some silliness.

"Heads up!" I sang out between chuckles, sailing his other shoe to shore.

Chuck had a thing about shoes—he was always striving for the perfect foot gear. After nearly freezing his feet on a ride with Milt after some bulls, he sent away for a pair of "bunny boots". He loved them because they kept his feet warm on the coldest days. But they weighed 7 pounds apiece, so he brought tennis shoes to wear when he got tired of walking in such heavy boots. The rest of us wore gum boots and two pair of socks.

By the time we staggered up the rocky trail with our supplies, Milt and Randall had the slaughterhouse open and the generator started. While Milt checked the compressors and cool rooms, Chuck and Randall moved the steers from the outside corral to the holding pen and I started the fires, one to heat water for the meat-cutting rooms and one in our living quarters.

Inside, we all had our jobs. Milt did the actual cutting, while the boys and I skinned. When a carcass was quartered, I washed it down and rolled it into the cool room on an overhead rail. But I wasn't big enough to wrestle the heaviest pieces of meat.

I kept the fires stoked, brought hot drinks from the bunk room, and, under the pretext of checking on the boat, dashed outside every 30 minutes to monitor the water, hoping against hope that the ice would move out before we finished.

As Milt cut, the boys watched.

"Yuck!" Randall screwed up his face at the sight of a beef tongue. "I can't believe people eat that."

I scrubbed its pebbled surface. "After I boil this with pickling spices and slice it paper thin, you'll gobble it up," I teased, knowing he'd be suspicious of everything I cooked for the next few days.

"Not me!" he snorted.

"Hey, lots of people eat this stuff," Chuck said, pointing to a beef heart. "Don't they, Mom?"

"Sure," I agreed. "Big-city delicatessens call heart, liver, brains, tongue, even kidneys 'variety meats', and use them in specialty sandwiches."

"Kidneys?" Even Chuck's eyebrows shot up.

"Well, maybe not for sandwiches," Milt broke in. "But the English make steak and kidney pie. My mother fixed it for me when I was a kid."

"I don't care," Randall insisted. "I won't eat it…and I'll be able to tell, so don't try to sneak it on my plate."

Milt grinned. "My mom couldn't fool me, either."

I made my rounds while they finished up, adding coal to the fires and checking the boat. Because our butchering facilities took up only one small area in a cavernous warehouse left by U.S. soldiers after World War II, it took a minute to walk to our bunk room near the entrance.

Outside, the sky was gray as flint and the water reflected back the somber color. It lapped against the beach, making soft sloshing sounds as it moved around the boat. I looked to the horizon and saw another boat's lights appear around Observatory Point a mile away, gliding toward the dock on the opposite side of Mutton Cove.

I knew Mike was already anchored up in the head of the bay and he was the only one scheduled, so this boat was a surprise. I couldn't make out the name, but the sleek white

SURF'S UP! When a storm blows in from the northwest, Chernofski Harbor no longer offers the safe refuge it does on calm days. The weather controls the lives of the Holmes family, and they've learned to respect the sea on those days when nature regains control.

shape looked familiar. I hurried back to tell the others. They were already working on the third animal. "The *Intrepid* just tied up at the pot dock," I announced.

Chuck's head jerked up. "Oh, no, not Sigmund! Are you sure?" He dropped his knife and bolted for the door.

I took his place, glancing across at Randall and Milt, who were both smiling. "Why is Chuck so upset?" I asked.

Milt shrugged. "You know Sigmund. He doesn't think Chuck is careful enough with his pots."

I really didn't know Sigmund. Like most of the fishermen who stored their crab pots with us, he was simply a voice on the radio, one of several with strong Norwegian accents. I'd never met him in person. He kept Milt supplied with some dreadful Norwegian goat cheese, which I'm sure was a great delicacy and hard to part with as far away from home as Sigmund was.

He also gave us crab, still alive and kicking from his tanks. With a king crab fetching as much as $15, it was a generous gift and one we loved getting. Ranch work kept us too busy to fish for ourselves, and we didn't have a big enough boat or any of the expensive equipment it took. I didn't blame Sigmund for worrying about how his equipment was handled. I hoped Chuck gave those pots all his attention.

The heavy door behind us burst open. "It's him, all right!" Chuck's face was a mixture of consternation and despair. He jammed his hands into his pockets and rocked back and forth on his heels.

"Vhat you doon, boy?" He mimicked the voice I'd heard on the radio with surprising accuracy. "You vatch out, vhat you do. Don't rip da veb on my pots vith dat backhoe fork."

Chuck jerked his hands free and locked them together. "If he says one thing to me—*Pow*, right in the kisser!"

"You know, Chuck," Milt said, turning on the electric hoist and lifting a carcass off a cart, "all you have to do is be careful with the man's pots and he won't yell at you."

"I *am* careful," Chuck defended, retrieving his knife and nudging me out of his way. "Accidents can happen to any-one."

"What did you do?" I asked.

LOADING POTS. Crabs can be big business for fishermen, and storing the crab pots they use provides an income for the Holmeses. The pots, really traps, are big, weighing between 600 and 800 pounds, and care must be taken in their handling, as Chuck found out from one hot-tempered fishing boat captain.

"I turned and two of his pots fell off the trailer. When I put them back on, the prongs got caught in the webbing. It didn't tear, just stretched it a little...of course, Sigmund had to see me do it."

"I know he gets pretty excited, Chuck," Milt said, "but he's a good customer and we want to do our best for him."

"I suppose," Chuck grumbled, "but does he have to call me 'Boy'?"

"You're lucky," Milt chuckled. "You should hear what he calls his crew."

Chuck held a carcass steady as Milt sawed it in half. "Shall I go see what he wants?" Chuck asked after the noise of the saw died away.

"He'll want the rest of his pots," Milt said. "The season isn't finished yet. You might as well go on over with the backhoe and get started on them." Milt looked around at the six quarters still waiting to be washed. "I'll come as soon as I can."

Chuck struggled out of his butchering gear, pulling it down over the bunny boots, then clomped away in those balloon-sized shoes.

"You go, too," I told Milt. "I can finish here."

"You're sure?"

I nodded. He didn't need any persuading. Crab pot storage was important to us.

"Yippee!" Randall yelled. "You promised I could drive the Bobcat!"

"That's right," Milt said. "I'll take the boat." Milt pulled off his butchering gear. "Let's go."

As their engine sounds faded, I started washing down the quarters. They were so big, each one weighing more than I did, and clumsy for me to move, even when hanging on rollers.

Before long, the entire room was wreathed in steam from the hot water I was spraying. The swaying carcasses looked like ghostly sentinels emerging from the fog misting up around my face as I worked.

When I finished washing, I took a sharp knife and started trimming. I was working on the last quarter when two men walked into the room. One was George Grunholt, skipper of the

Akutan, a processor anchored in the harbor. With him was a dark young man who stared at me with wide eyes.

I must have looked strange to him, with knives and meat trimmings strewn everywhere and the ghostly steam floating all around me, like something out of a bad horror movie. I laid down my knife and gave them my best smile. "Hi."

George, whom I had met briefly before, answered. "I'm looking for the water Milt told me I could get at the valve down on the dock. Our water maker is on the fritz."

"It's turned off right now," I answered. "I need all the water pressure I can get in here." I opened the cool room door and started pushing the unwieldy quarters along the rail toward it. The young man jumped hastily back. "I'm almost done," I said. "Then I'll show you where to turn it back on."

"Can I help?" George asked.

"Sure," I pointed through the door, not wanting them to touch the freshly washed meat I was handling with gloved hands. "Go pour us all some coffee." I gave him directions to the bunk room. "Make yourselves at home; I'll come as soon as I finish."

The bunk room, with its sawhorse table and rusty iron beds, exuded a friendly warmth after the chilly atmosphere of the butchering rooms. We huddled around the potbellied stove and warmed our hands with mugs of thick boiled coffee.

George was part Aleut, a native of Sand Point, Alaska. He was about 60 years old and had been Milt's friend for many years. I first met George when he brought the *Akutan* into the harbor in November to buy crab from fishermen and process it on the ship.

His crew was mainly young people from the Seattle area, a lot of them college kids earning money for school. Some were Asian immigrants new to America and still groping with the language—like the young man beside George at the moment. I didn't catch the man's name when George introduced us, only that he was from Vietnam.

When I extended my hand, he pumped it enthusiastically while an excited volley burst from his lips. I smiled and nodded without understanding. I thought he was thanking me for the coffee.

After I showed the men where the outside water valve was, I walked out on the dock. A cold northwest breeze touched my face. Good, I thought, it will blow the ice out of the channel. Tiny riffles scudded across the water, and beneath my feet I heard the gentle splash of waves washing around the pilings.

Directly across the bay to the south, I could see Cutter Point jutting into the water. The sight of it always gave me comfort—only a mile distant and protected from the north wind, it was the nearest landing on the headquarters side of the harbor.

It was a good mile and a half away from the house, but if the wind ever got too bad while we crossed the harbor, we could duck behind Cutter Point, leave the boat and walk home. I started back down the dock, thinking about the mess I still had to clean up.

"Mom! Mom!" Randall yelled, dashing around the side of the slaughterhouse. His head was bare, his gloves were gone and his open coat flapped behind him with each piston stroke of his legs.

"What's the matter?" I sprinted the remaining yards.

"There's another boat at the dock," he gasped, his brown eyes alive with excitement. "The *Teacher's Pet.*"

"You ran a whole mile to tell me that?" I pulled his coat together and managed the top button before he jerked away. "What's so special about it?"

"Miss Universe is on it!" He stopped to catch his breath. "No, not Miss Universe...Miss Iceland...or something like that."

"Now, Randall," I chided, knowing his penchant for exaggeration. "Are you sure?" His enthusiasm sometimes distorted the facts, but his stories always contained an element of truth. I probed. "Who told you that?"

"Helga."

I raised my eyebrows.

"That's her name," he insisted. "She don't talk English too good, but she's real pretty and about 10 feet tall." He stretched his arm above his head. "And she asked to meet you; she wants some sheep heads."

His words tumbled over one another. "They eat them! Can you believe that?" He made a gagging sound. "I might eat tongue if I was starving to death, but never a sheep head!"

This had gone far enough. "Randall, what kind of dumb joke is this? Nobody eats sheep heads." I walked into the long warehouse to the meat cutting rooms. He followed me.

"It's the truth!" he declared. "Honest. Come see for yourself."

"Okay," I told him. "But first we have to clean up this mess."

For once he didn't complain. While I washed the equipment, he hosed the floor, then together we wrestled the heavy wheelbarrow of scraps outside. The eagles and foxes would make short work of them.

Before we left, I found him some gloves and made him put on his stocking cap. As we walked, the wind was in our faces, brisker now and very cold. I noticed the erratic tire tracks made by the front end loader. "Did you have any trouble with the Bobcat?" I asked.

"Nah." He turned and faced me walking backward into the wind. "Milt said I did good."

"Did Sigmund say how many pots he wanted?" I asked Randall. Milton and Chuck could move 48 pots an hour, and I hoped they would finish by dark. The increasing wind brought with it a niggling unease. The bay could become a furious caldron of smoking water within minutes if this rising wind were part of a storm front. I hadn't listened to Peggy Dyson's 8 a.m. weather on the marine band radio before we left home because I was too busy. Now I wished I had taken the time.

"I don't know how many pots they're loading," Randall said, then grinned sheepishly. "I watched TV while they talked. The cook gave me some ice cream."

"It doesn't matter," I said. "I'll ask Milt when we get there."

We walked as fast as the frozen ruts would allow, and as we got closer, I saw Chuck pulling a loaded trailer along the beach road from the gravel point where we stored the pots. That meant Milt had the Bobcat on the point another half mile away.

When we reached the dock, Randall grabbed my arm and

pointed. "See, just like I said, the *Teacher's Pet.*"

A white boat, beautifully painted, with the name etched prominently on the bow, was tied up next to the *Intrepid.* "Come on, Mom." Randall tugged my sleeve.

"I want to talk to Milt first," I told him. "You go ahead."

Randall dropped my arm and scampered across the plank bridging the boat and dock. In a moment, he disappeared from sight.

Chuck was waiting for the boat crew to transfer the pots from his trailer to their deck when I passed. Instead of speaking like I had planned, I just waved. He acknowledged me with a slight nod and turned back to the three young women clustered around him. I guess Randall hadn't had time to pass the word about Chuck's age.

I found Milt stacking pots with the Bobcat onto our second trailer. He jumped out of the cab when he saw me. "How much longer?" I asked.

"A couple hours," he said. "We'll have to come back tomorrow. I still have Mike to do." He rubbed his hands together. "I'll have to chill down the cool room again, too."

I unzipped my coat and put his hands inside. "Man, it's cold," he muttered. "This wind must be blowing off a glacier in Russia."

Chuck clattered up behind us with an empty trailer. Milt pulled his gloves on and went to unhitch it. I watched for a minute as they jockeyed the vehicles around, exchanging the loaded trailer for the empty one. When they finished, I jumped on the trailer and rode back to the dock, wondering again where Randall had gotten his story and what to expect when I met his "Helga".

The three women who'd been talking to Chuck, little more than girls, sat on an overturned fuel drum. I waved as I passed them on my way to the *Teacher's Pet.* "Hi," one called. "Are you going to see Miss Iceland?"

I gave them a surprised nod and hurried on. So it was true. What in the world was Miss Iceland doing on a fishing boat in the Bering Sea?

I got my answer soon enough. When I crawled over the rail

on the boat's back deck, a huge blond man ducked out the galley door. When he straightened up, he towered above me. "Ja?" he asked in a soft voice I could not believe came from that massive chest.

"I'm looking for Helga," I faltered.

A wide smile split his face. He motioned me through the door. Three steps down and I came face-to-face with one of the most beautiful women I'd ever seen.

While not quite 10 feet as Randall had described, she was indeed tall. Tall, willowy and exquisite. She was bent over the minuscule galley sink, peeling potatoes, her silvery blonde hair carelessly pulled back with a rubber band.

"My vife, Helga," the man behind me said. "Miss Iceland and second runner-up for Miss Universe 1972." The pride in his soft voice spoke louder than a shout.

The woman lifted her head. "Gunnar." Her clear blue eyes pleaded. Then she looked at me and smiled. "He embarrass me." Wiping her hand on her sleeve, she held it out. "You verk on boat?"

NORDIC BEAUTY. There really was a Miss Iceland on the fishing boat, as Cora found out when she met Helga and her husband, Gunnar. They visited the ranch so Helga could ride horses with Randall, as he promised, and to procure some sheep heads, which this couple considered a delicacy.

"No," I shook my head. "I'm Mrs. Milt…" My words were cut off as I was enveloped in a bone-crushing hug.

"Milt's new vife!" Gunnar said as he lifted me off my feet and kissed me on both cheeks. "Ja, gut!" He dumped me onto a galley bench and sat beside me. "From da sheep ranch," he said to his wife. "Dey haf sheep heads."

Helga's eyes lit up. "Randall?"

"My son," I nodded. "He said you wanted sheep heads."

"Ja." She set the potatoes on the stove. "Gut vith spuds."

"I'm sure," I said lamely. "How do you fix them?"

With a great deal of relish and stumbling broken English, she explained in detail how to cook a sheep's head…she even patted her stomach.

"We're butchering sheep soon," I told her. "I'll ask Milt to save the heads."

"Tanks." She brought coffee to the table, sliding into the galley seat opposite us. "Randall haf horse?"

"Yes," I said, wondering what that had to do with sheep heads. "George."

"Ja, George." Helga beamed. "He say he take me ride on George." She pantomimed pulling back on the reins. "I luf horses. I rode all time in Iceland." A wistful tone crept into her voice.

"Do you miss Iceland?" I asked.

She shrugged, "Ja, sometimes." She stared out the porthole at the brown grass-covered hills. "But I like here; no trees—just like home, only little booshes."

"Why did you leave Iceland?"

She looked at the fair giant sitting beside me. "My husband is fisherman," she replied softly. "So I live on boat."

I sipped my coffee and looked at her. Later I learned she had modeled for *Cosmopolitan* and was a ballerina with the Reykjavik Ballet for 10 years. But for that moment, she—tall and beautiful in bib overalls, with wind-roughened cheeks and chapped lips—and I—short and plain, with dried blood on my face and under my fingernails—were just two women who had followed their hearts.

We smiled at each other.

ISOLATED BEAUTY. For all its desolation and cloudy days, it takes only a day of sunshine to reveal the stark beauty of Unalaska Island and Chernofski Harbor...and to show why Cora, Milt and the boys love this land. Such days are rare and give Cora, Milt, Randall and Chuck all the more reason to appreciate where they live.

Chapter Five

WINTER WATER

As I crawled on hands and knees across the plank connecting the *Teacher's Pet* to the dock, I saw both the backhoe and the Bobcat parked on the road. The wind whistled through the boat's rigging above me. I heard water slapping against the pilings and felt the boat sway and grate along the dock with each wave.

Chuck saw me and broke away from a small group of people on the beach. They were processors from the *Akutan*, waiting for the skiff to come from the ship and take them back. I recognized the three girls he'd been talking to earlier and the young Vietnamese man George had introduced. He was talking and gesticulating with much more animation than he'd displayed at the slaughterhouse.

When Chuck got nearer, I saw him laugh and shake his head. "You want a ride, Mom? Milt just left in the boat; he thought you had walked."

"Sure." I stepped around some rotten decking. "What's so funny?"

"You know what that guy is saying?" he snickered. "'A little lady butchers beef all by herself.' He's convinced everyone you're some kind of wonder woman."

"Me!" It was preposterous; I could hardly pick up one hoof.

"Yeah. Ha, ha, ha, ha." Chuck doubled over. "Hee, hee, hee, hee."

"Okay," I said. "It's not *that* funny. Did you set him straight?"

"No." Chuck wiped his eyes. "I think it's great." He strutted beside me in his bunny boots, flexing his arms above his head like Tarzan, laughter erupting from his chest. "My mom, The Butcher of Chernofski."

"Where's Randall?" I asked, hoping he hadn't heard the story yet, or I'd never hear the last of it. "Don't tell him, okay?"

"Tell me what?" Running footsteps sounded behind us. "Nothing," I looked a warning at Chuck.

"Come on, Chuck, tell me," Randall cajoled as they jogged off. Before they reached the parked vehicles, I heard fresh laughter ring out. I sighed. Randall could pester a secret out of a rock.

I climbed up beside Chuck's high seat and tried to maintain a hurt, dignified silence, but I could never stay mad at either of them for long. And I had to admit, it was too good a joke to keep.

Milt was shutting down the generator and checking the temperature in the cool room when we got there. As soon as the boys parked their equipment, we all got into our raingear and piled into the boat.

"I need to stop at the *Akutan* and talk to George for a minute," Milt said as he jerked the outboard to life.

Our heavy wooden dory plowed through the waves. Still apprehensive about ice, I scanned the water as far as I could see—no ice in sight, but the channel looked rough, with a white, curling chop on it, and the north wind blew harder and colder across the open water.

I pulled my cap over my ears and hunkered down on the seat beside Milt. Randall crawled up under the bow and Chuck sat facing the wind, enjoying every drop of salt spray and the crazy rocking of the boat.

When we reached the *Akutan* a half mile offshore, Chuck threw a line around its rail, scampered onto our bow and held the dory alongside the ship so Milt could get aboard. Next, Chuck tied us alongside with the double half hitch Milt had taught him.

Gloves held in his teeth, Chuck's fingers grappled with the wet stiff line. "He loves boats," I thought with a pang, "and I don't want him to." Every week we heard "Maydays" on the marine radio; sometimes boats sank and whole crews perished—hardly more than kids, some of them. When I listened to the Coast Guard listing names and ages, I would weep for the anguish their loss caused. My lips tightened on a soundless wail. I didn't want Chuck to be a fisherman.

Under the bow, Randall didn't move. He'd get seasick at

the first hint of roughness, especially if we weren't moving, so I knew he was fighting nausea. Sympathetic toward his discomfort, I was nonetheless glad he had no desire for a life on the water.

Milt's minute turned into 10 before he returned, and all the time the wind got steadily stronger. Our dory slammed against the *Akutan* and bounced up and down with each new wave. Randall crawled out from under the bow and

CORA AT THE TILLER. Cora readily admits she's not much of a sailor, except on calm days like this. On one stormy trip, she had Milt beach the dory so she could walk home, rather than suffer any more on the rolling waves.

gulped in deep breaths of the raw air just as Milt appeared. "Hey, Randy, someone wants to talk to you."

On deck stood one of the girls I'd seen at the dock. She was a small blonde, perhaps 18, pretty, with a friendly smile. She looked at Randall's pinched face uncertainly. "Uh, are you Randall?"

"Yeah," he gave her a watery grin.

She shifted from one foot to the other. "Um, I was talking to that woman on the *Teacher's Pet...*"

"Helga," Randall prompted.

"Uh huh, and...she said you had a horse?"

"Yeah," Randall answered, flinching as the boat swooped down and then up again. "George. Why?"

Her words came out in a rush. "If I came over to the ranch, would you take me for a ride on him?"

Despite Randall's misery, he flashed a big grin. "You bet."

Delight lit up her face. "Other girls want to come, too, okay?"

"No problem," Randall said grandly. "Bring as many as you want."

"Thanks," she waved and disappeared inside.

Randall, triumph written all over his green features, turned to Chuck, who had listened to the entire exchange with an air of disbelief, and thumbed his nose. Randall didn't need a mustache...he had George!

Our outboard roared to life and Chuck cast off the line. The minute we moved away from the *Akutan's* protection, the wind caught us. Suddenly we were being thrown every which way. A green wall of water rose up in front of us while the boat shuddered and broached sideways into the trough. As we plowed into the next oncoming wave, the spray hit us like it was coming out of a hose. I landed on the floor between the seats. I heard Milt shouting for Chuck to bail.

It was even worse when we reached the unprotected channel. We were tipping over! A new wall of water hit us every 20 seconds. I slid between the seats like a sack of potatoes. Water

WOOL GATHERING. This boat is docking to pick up the wool clipped from the Chernofski flock. The round burlap bag in the foreground is full of wool and weighs about 350 pounds. Cora kept some of the wool for herself and hoped to make clothing from it when she finally conquered her spinning wheel.

sloshed over my knees as they banged into the sides of the boat.

Where were the life preservers? Randall was lying on them under the bow. Why hadn't I insisted the boys put them on ...and Milt? Not that they would help. Even a strong swimmer would freeze before he reached the shore.

Surely no boat could withstand such a beating for long. Above my head I heard Chuck's excited voice. "Here comes a big one!" he screamed. The bow went down and we swerved sideways with a sickening lurch.

I threw myself across the plank seat and pressed my face against the slippery wood. Salt water burned my eyes and stung my lips.

"If it doesn't get any worse than this, I can stand it," I whispered into the maelstrom, using the litany that worked in dentist chairs and traffic jams. "If it doesn't get any worse than this, I can stand it. If it doesn't get any worse than..." Another wave broke over the bow. Water cascaded over me. I looked around wildly.

"Take me to Cutter Point!" I shrieked.

Milt glanced down from his seat where he was fighting with both hands to keep the bow pointed into the waves. Water poured off his hood. He made a negative motion with his head. "Now!" I shouted. "Right now!"

He took a closer look at my terrified face and changed course. The boat came around, rolling off the wave until the foam-flecked water was level with the rail.

It was too much. I got sick, bending over the side, the water 2 inches from my chin as my braids dipped into the sea.

The moon had risen above the hill; full and bright, it illuminated the mountainous crests chasing us toward shore.

The sea was coming up behind us! I felt the boat raise with a long slow motion as a wave overtook us. We nosed off the crest and fell into the trough between waves with a jarring crash. Before I could grab hold, we were up on the next wave.

Then came another slow raise followed by the abrupt downward plunge, but no more broaching sideways and wallowing in the troughs; no more water slapping over the rail. I crouched on the floor, peering into the spray, watching for shore.

BEACHED WOOD. When storms like this blow, boats are not safe at sea or in the dock. The logs that blew ashore often endangered the marine ways, where the ranch dory was kept. That meant Cora, Milt and the boys often had to stand guard for wayward logs until the storm stopped.

The boat veered sharply to the right and suddenly we were there. Milt slowed the engine and we idled smoothly up beside the Cutter Point dock. I leaped out and stumbled over driftwood logs on the debris-cluttered beach.

Alive! All of us were still alive! I clutched handfuls of brown beach grass and held on to the solid earth. Words of thanks tumbled from my lips between sobs and gasps. Nothing felt as good as that cold, wet earth beneath me.

As soon as they had the boat tied off, the boys and Milt climbed up the beach. "Hey, why did you make Milt change course?" Chuck glared at me. Behind him Randall looked just as murderous.

"I didn't want us to drown," I said weakly.

"You chicken," Chuck scoffed. "It wasn't bad." He looked at Milt. "Was it?"

"We weren't in any danger," Milt said with a noncommittal shrug. "But I don't want your mom scared, either." He helped me to my feet.

"It won't hurt us to walk home," I breezed, so grateful to be off the boat I could have walked clear to town.

"Oh, sure," Chuck retorted. "You're not wearing 7-pound boots on each foot." His cheeks were bright red. His lips moved furiously. "You...you...you overreact to everything!" He whirled around and stomped off.

"Yeah, Mom," Randall chimed in, his face still paste-colored. "It was no big deal." He hurried after Chuck.

The wind blew in our faces and the full moon lighted our path. Beside me Milt said, "I would never take you or the boys on the water if I thought it was the least bit dangerous."

"And you?" I asked.

"I've lived here 23 years and I've been scared a couple times." He helped me over a log. "But don't worry about me—I'm careful." He hesitated. "And so is Chuck. Careful and cautious." He didn't mention Randall. Randall wasn't allowed to drive the boat by himself, yet. I was grateful for that.

When Milt and I reached home, the boys had already gone to bed. The fire was long out and the house was like an icebox. We didn't even turn on the power, just washed up in lukewarm water and went to bed ourselves.

All night the wind screamed around the house. It sounded like a thousand circling jets. The walls shook with each gust. Sleet and ice pellets hammered the windows. Listening to the storm and Milt's even breathing, I envied his ability to sleep through such a cacophony. Would I be used to it when I'd lived here 23 years as he had? Never! I was still awake when the alarm sounded at 6 a.m.

Storm or not, I knew Milt would go across the harbor even if he had to ride around 8 miles of coastline on horseback. I crawled out of bed without waking him and groped my way to the kitchen.

Most of the time I loved the hulking monster stove, but not when I had to start it in the cold and the dark. I always got coal soot on me raking clinkers and ash from the firebox, and the noise usually woke everybody, especially Chuck, and I didn't want him up until I had the fire lit. Splitting kindling was his chore, and he always griped at how much I used.

From the four sheets of rationed newspaper, crumpled diagonally and placed under 20 sticks of cedar kindling and 15 pounds of hand-placed coal, to the 6 feet of waxed paper shoved up the flue, I laid that fire with all the attention to detail I had used in starting IV's in sick babies, because I knew I only had one chance. If I didn't do it right, the paper and kindling would burn up without igniting the coal and I'd have to dig all the coal out and start over. I really didn't like doing that.

When it was laid and the drafts all open and double-checked, I struck a match and held it to the paper under the kindling. As soon as it took off, I lit the waxed paper. I heard a satisfying *whoosh* as fire rushed up the stovepipe, creating a tremendous draft, and the cedar kindling started to crackle. Feeling a sense of triumph as keen as if I had resuscitated a choking baby, I closed the firebox door. I liked doing a good job no matter what it was.

"I'm sorry I yelled at you last night, Mom."

I whirled around. Chuck stood in the doorway just outside the circle of lamplight.

"That's okay," I shrugged. "Any blisters?"

"Nah, just sore." He moved down the length of the kitchen and peered out the plate glass window fronting the bay. He cupped his hands around his eyes and pressed them to the glass. "At least it stopped hailing."

"But I think the wind is worse," I said. "Would you get some bacon out of the storeroom for me? I don't think I can get the door open."

When he returned, I was mixing powdered milk. He looked at it and made a wry face. "I wish Bossy hadn't died."

"Me, too," I agreed, slapping bacon in a cast-iron pan. "Too bad none of us were home when she bloated. Milt might have saved her."

"I even miss milking her," Chuck said, clattering dishes onto the table.

"Well," I said, "her calf, 'Tulip', will come fresh next spring." I folded soda and sugar into my sourdough pancake batter with a wooden spoon. "We can try milking her even if she is mostly Hereford. She'll have a little extra, but it won't be as much as Bossy gave."

DAIRY STORE. Milk cows like Tulip (above) are a valuable asset to the ranch because fresh milk and butter are hard to come by when you order groceries once a year. Peep-Sheep, the "bum lamb" of the ranch, could be a handful at times, but everyone treated this pet like one of the family.

Indeed, we missed Bossy sorely. Gentle and generous, she gave us $15 each day in dairy products—at Alaskan prices. With milk at $6 a gallon, and butter over $4 per pound, both nearly impossible for us to get fresh, we didn't realize how fortunate we were until she died. And I was just getting good at making cheese, too!

I sprinkled oatmeal into the pan of simmering raisins and walnuts, pulled it to the back of the stove and went to wake up Milton and Randall.

As I set the lamp on our dresser, a violent gust of wind shook the house. "Thank God we left the boat behind Cutter Point last night," Milt said as he threw back his covers. "We'd never get her off the front beach this morning."

"Can't you radio Sigmund and tell him it's too rough to go across today? I don't want you walking back over there on your bad leg." Milt pulled me into his arms for a quick hug. "Don't worry," he whispered into my hair. "We'll take the new track machine; Chuck loves to drive it." He released me and limped into the bathroom, his uneven footsteps loud on the bare hardwood floor.

When we sat down at the table, Randall asked Milt, "Does Mike have our mail?"

"No," Milt told him. "I talked to him on Sigmund's radio yesterday and he said that when he asked at the post office, someone else had already picked it up for us."

"Uh oh," I broke in. "I hope they didn't change their mind about coming this way."

"My traps are never going to get here," Randall moaned.

"I don't care if the mail never comes," Chuck said. "I know my second semester lessons will be in it."

Chuck had been enjoying a long break. Since he was in high school, his correspondence courses came by the semester, and only one semester at a time. No amount of explaining our lack of mail service would change the rules. So Chuck hadn't had classes since he finished his first courses 6 weeks ago. Randall, on the other hand, was in fifth grade and his lessons for the whole year came all at once, much to his disgust.

"If I don't get those armature brushes for the generator soon, we might be in the dark," Milt warned, carrying his dishes to the sink. "I ordered them in September." He grabbed his coat from its hook behind the stove.

I hope I have a letter from Mom, I thought as I started washing dishes. It's been months since I've heard from anyone in Idaho. The last time we had mail was before Thanksgiving, and here it was 2 weeks past Christmas. Maybe our Christmas orders will be in it, I thought.

The outside door slammed and brought me out of my daydream. For a few minutes, I had forgotten the wind.

"We're off," Milt spoke from the doorway. "I'll call you from Sigmund's boat." He followed Chuck out into the storm.

Still at the table, Randall glumly mopped up the last of his syrup with a pancake. "School day, I guess, huh?"

"Good guess." I pressed my face to the window, watching the tiny pinpoint of light crawl slowly up Cutter Point hill and disappear over the top. "Just as soon as you feed the chickens and let out the dogs."

With a sigh that came from the bottom of his chest, he dragged his chair back. "I hate school," he grumbled. "It's always the same old thing."

"Not today." I forced a cheerful tone into my words. "You get to take a science test and learn how to conjugate verbs."

"Conju...what?" He leaned his chin on his hand in a pose of total boredom.

"See," I said brightly. "It'll be new and exciting. Now go do chores while I finish the dishes." As I turned away from the window, a light appeared around the headland, bobbing crazily up and down on the bay. "A boat is going out," I said. "The wind must be letting up."

"Don't sound like it to me," Randall said, struggling into his heavy coat. Just then Sigmund's voice crackled over the radio, calling for Milt.

I hurried into the radio room, but before I could answer Sigmund, another voice broke in, loud and disapproving. "Dis is no time to be asking a man to come across da harbor. I'm taking vaves over da vheelhouse."

I thought about the light on the bay, and then it struck me...I clutched the microphone and screamed. "He's already gone!"

No one answered. Into the dead silence, Randall shouted from the kitchen. "Hey, the boat is turning around! It's too rough out there for him."

I remembered the *Silver Clipper.* "Mike!" I shrieked into the microphone. "Milt and Chuck are out there! Can you see them?"

In a second, Mike's gravelly voice came on the air. "No, Cora. When did they leave?"

"They drove to Cutter Point on the Ferret about an hour ago." My voice cracked. "Please, Mike, find them," I begged.

"What's going on?" Another deep male voice broke in.

Someone else answered. "Two guys from the sheep ranch are coming across the bay in a dory."

"Good grief! In this?"

My knees buckled. I sank blindly onto the stool behind me. Then a high thin voice wavering with excitement shouted.

"I see them! They're bouncing out of the water but they're..." He stopped abruptly. "A log! *They're going to hit it!*"

I put my head between my knees, wondering, in a strange detached way, why my skin felt sweaty. My hands were so slippery I could hardly hold the microphone. I stared at it in numb helplessness, hearing the sound of impact, splintering wood, seeing the water rush in, hearing their cries. "It's happening," I thought dully. "It's really happening."

From far away, I heard Mike's calm voice. "They missed it." He spoke soothingly. "They're going past me now."

I couldn't answer.

"I have them in my light!" someone yelled.

"That's Brad, skipper on the *Polar Command*!" Randall hollered from the doorway. "They're almost across."

"Ja, gut," came Gunnar's soft accent. "Dey safe, Mrs. Milt. I see da boat on shore."

"Thank you," I whispered past the ache in my throat as my eyes filled and spilled over. "Thank you, all."

Chapter Six

THE FLY DIED

"Mom, you're singing again!" Randall looked up from his books with a pained expression.

"Oops, sorry." I stopped humming and finished putting away the breakfast dishes without slamming cupboard doors. Randall liked quiet in his classroom and I usually stayed silent as I did my kitchen chores. But I couldn't help wanting to rejoice—Milt and Chuck were safe.

"Now you're whistling!" he complained 10 minutes later.

"I forgot. Sorry." I clamped my lips together and gave the bread dough I was kneading a final punch.

"I can't concentrate when you're making so much noise," he grumbled. "Why don't you spin or something?"

"Oh, sure," I laughed. "I'd be screaming and cussing so loud you'd never get any studying done."

"Yeah," he grinned. "And you'd ask me to help you turn the wheel and I'd get out of school."

"Maybe later."

I covered the dough and set it on the warming oven to raise. "Have you read that history assignment?"

He sighed and slouched over the book. "Who's this guy?"

I leaned over his shoulder. "Attila the Hun," I said. "He lived in the 5th century."

"What does 'scourge' mean?" he asked.

"Someone really mean," I answered, reading the page. "He was called the Scourge of God because he was so fierce and he tried to conquer Europe."

I racked my brain thinking of something to make the shadowy historical figure come to life for him. "He was a small man, not even 5 feet tall." I measured above my own head to give him an idea. "He wore one of those battle helmets with horns on it, maybe to make him look taller."

SCHOOL'S OUT, or why would Randall be smiling? School for the boys was in the ranch house kitchen and their desk was the kitchen table. Cora was the teacher, using lessons that were shipped in by boat. Those were the few times the boys wished the mail was late!

"Okay, okay." Randall closed the book. "I just wanted to know who he was, not his life story."

"Well, that's who he was, a short, mean person who wanted to be boss of the world." I opened my folder and scanned the weekly agenda. "Are you ready for that science test?"

"I'll never be ready for that test," he groaned. "How do I know what's inside my own body? It's covered up with skin."

"You're supposed to read the chapter and look at the pictures," I scolded. "Did you do that?"

"I started, but it was boring so I read my trapping book instead." Immediately his face brightened. "You know, Mom, it tells how to get a fox out of a trap. You hit him on the nose ...doesn't even hurt..."

"Randall!" I gave him my sternest look. "If you don't do your lessons now, you'll never get into college."

He looked at me as if I'd grown another head and said in all seriousness, "I'm not going to college." He straightened his shoulders. "I'm not even going to high school, if I can help it."

"What do you plan to do when you grow up, then?" I fell back on that much used, trite question that's supposed to alert children to the looming responsibilities of adulthood. "How will you make a living?"

"I'll trap foxes," he said with that stubborn expression I knew so well. "And I'll break wild horses." His eyes glowed. "Milt promised to teach me to shear sheep, and I already know how to round up cattle." Then came the clincher, the one he always used in these battles we waged about education. "Milt never finished high school and look how smart he is."

His face radiated triumph. He had me there. Leaning back in his chair, he crossed his arms, waiting.

There was no point in explaining the difference between quitting school because you wanted to and because you had to. To Randall, the fact was all that counted.

Milt left school at 16 to tend his family's flocks in the Idaho backcountry. But his learning didn't stop. His saddlebags always held a book or two; thin well-thumbed volumes of Tennyson and Longfellow, books whose worn leather covers still lined the top bookshelf above our bed. Formally educated or not, Milt was the most learned man I'd ever met. And he had done it without benefit of high school or college.

Yet, on the other hand, my education had put food on the table for a lot of years and I was always grateful for the security it afforded me. How could I justify not urging Randall to take advantage of every scrap of learning that came his way? I couldn't.

"Well, he did finish fifth grade," I reminded Randall, ending the discussion by handing him his science book. "You have an hour to review."

While he studied, I made loaves out of my dough and returned it to the warming oven to rise a second time. I replenished the fire from the coal bucket beside the stove and wandered to the window.

The wind shrieked around the house. Salt spray, picked up from breakers smashing against the beach 300 feet away, spattered the glass. A draft curled around my legs. When I put my hand over the electric outlet under the window, I felt cold air push against my fingers. Even with the coal fire roaring up the chimney, this end of the room was chilly.

I looked out the window again, glad it wasn't raining, and hoped Milt and Chuck were warm enough…hoped one of the boats invited them aboard for lunch…prayed they wouldn't try to come home until the wind died down.

When the loaves were 2 inches above the pans, I slipped them in the oven, then sat down across the table from Randall and studied the next English lesson. We were learning about verbs, action words, and how to change them from present to past tense and then to past perfect.

That shouldn't be too hard to explain, I thought, searching my mind for examples. The wind blows today. It blew yesterday, and it has blown many times. I congratulated myself on their brevity and clarity.

I checked the bread, covering the loaf next to the firebox with tinfoil to keep it from burning. The wrinkled foil had been used so many times I had to pinch the tears together. In the back of my mind I jotted aluminum foil on my grocery list. After another 15 minutes, I tested the bread for doneness. The crust emitted a satisfying hollow echo to my tapping finger and I emptied the loaves out of the pans onto the counter to cool.

Randall snapped his book shut. "I'm starved." He sniffed the air. "Do we have to wait until Milt gets home to cut the bread?"

Getting the heel off a fresh-from-the-oven loaf of bread was a special and greatly coveted treat. Unless Milt was home, I wouldn't cut a warm loaf because it squashed it all out of shape. We were still newlyweds and I couldn't refuse him anything.

With the boys, I was more hard-hearted. But Randall could be a charmer, too. His usual method was to find Milt in the pastures or outbuildings and tell him bread was coming out of the oven. Since Milt was gone today, different measures were called for.

He came and stood beside me as I slathered bacon grease over the hot crust. He didn't say anything, just watched me anoint all four loaves and cover them with a towel. Then he sighed.

"Oh, all right," I gave in. "But let me cut it!" I yelled, grabbing for the bread knife he had hidden behind his back. A wide grin spread across his face as he watched me ever-so-gently saw off the crusty end. "Now that it's already cut, can I have two slices?"

"Don't push your luck," I warned. I cut him another thick slice and one for myself. While Randall made us cups of hot tea, I got out the jam.

We ate in companionable silence, but we didn't always. Lunch marked a time-out in our school day, a quiet hiatus from discussions, skirmishes and often head-on collisions. There were days, if our confrontations had been intense, when Randall spent his lunch hour running off his anger on the beach or flung across his bed with his head under his pillow, while I hid in the bathroom and wept tears of frustration.

Our personalities

HOMEMADE BREAD, the only kind at the ranch, was sure to stop lessons when the wonderful aroma rose from the oven. There wasn't always time for a leisurely lunch, but the bread was always fresh.

were different; he was stubborn and I was quick-tempered. The fact that there was a deep and genuine love between us didn't always mean we liked each other. And school was where we disliked each other the most. But I tried hard to be fair, and Randall never held a grudge.

When the hour was over, Randall brought me the book we were reading, then stretched out on the floor in front of the stove. He crossed his arms behind his head and closed his eyes. This was his favorite part of the school day, the half hour I read to him out of a book of his choosing. We were reading *The Keep*, a suspense story set in Transylvania, and Randall loved it. After 3 days of no classes, he was more than ready to hear the next chapter and begged me not to stop when it was finished.

"Sorry," I marked the place and closed the book. "If you want more today, you'll have to read it. But not now; it's science test time."

"Let's get it over with, then." I got him the single Xeroxed sheet and he settled himself on the open oven door and read it with the resigned air of a martyr. "Oh, great, name five body organs. I'm dead."

While he was doing penance, I punished myself by sitting at my spinning wheel. For a minute, I just looked at it. It was beautiful, with its satiny maple finish. And easy to use, with its simple flyer and bobbin arrangement. So why couldn't I make the darn thing work?

With a sigh as despondent as any Randall could utter, I opened Candace Crockett's *Complete Book of Spinning*. Before I could read one word, Randall erupted off the oven door.

"Variety meats!" he shouted. "Organs are variety meats!" He danced around the kitchen waving the sheet of paper. "The heart is an organ!" He stopped and wrote madly. "The liver is an organ!" He kissed the paper.

"Oh, yeah…and kidneys! Milt said his mom tried to make him eat them." He squeezed his eyes shut and clapped his hand to his forehead. "What else was in that steer? I need two more." He glanced at me, but I shrugged; he was doing fine on his own.

I hid a million-dollar smile. Randall might not think school was important, and he might think he hated it, but nothing could stop him from learning. He remembered everything he saw and heard outside the classroom. He was bright and quick, but nothing would ever make him a scholar.

"I'll guess lungs and tongue." He scribbled the last two answers and handed me the paper with a flourish. "How'd I do?"

I looked over his answers and nodded. "You passed, I think."

"That's all I care about," he said. "Put it in the mail."

Randall's correspondence program was part of Alaska's regular school system and he had a teacher in Juneau who graded all his papers and made comments. My only responsibility was to see that he did the lessons. But I made sure all his assignments had passing grades before I let him mail them so he didn't waste time repeating lessons.

"Great," I said. "All you have left is English and math and you're through for the day." I slid my chair to the table. "And English is easy for once."

Opening up my lesson plan folder, I explained to him, using my examples, how to conjugate verbs. "Now you do one," I suggested, wanting to be sure he understood before he tackled the workbook assignment. I looked up and saw ribbons of seaweed flying past the window.

"How about fly," I said. "Give me a sentence using the past tense of the verb 'fly'."

He thought for a minute and said, "The fly died."

I just stared at him, unable to speak. I wanted to laugh and cry at the same time. It was obvious he had no idea what I was talking about, yet he had quickly come up with a sentence dealing with fly and past tense. A dead fly was certainly past tense. I admired his inventiveness.

"Good try," I said. "Maybe I can explain it better with an..."

"Chernofski. Chernofski." Milt's voice boomed over the radio. I hurried to answer. "Chernofski back. What's going on, Milt?"

"Lunch on Sig's boat," he told me. "The wind's still whipping it up over here so we'll stay in the slaughterhouse tonight, okay?"

"Good idea," I said. "Randall and I can handle chores."

"Well, you be careful going through gates and doors. This wind will take them away from you and knock you down with them."

"We'll go together," I promised. "How's Chuck?"

"Fine. He's doing a great job on the backhoe. We'll be finished with Sigmund this afternoon and start on Mike just as soon as he can get his boat alongside the dock."

"Tell him I said 'Hi'," I said. "And you take care of yourself.

I'll keep the radio on all night." I signed off without any endearments because every boat in the harbor was listening and I didn't want anyone coming back with rude kissing noises or sweet nothings uttered in a high falsetto, like I'd heard whenever someone got sentimental on the air.

Milt's call ended our lessons for the day. By the time I got back to the kitchen, Randall already had his wraps on. I looked at the clock—nearly 3 p.m. "Wait for me," I said. "I dumped the coal bucket into the fire and handed him the empty. "We'll fill these first."

The minute I opened the storm door, the wind grabbed it. Before I could let go, it jerked me outside. The door slammed into the side of the house with a crash. I threw my body against

ELECTRONIC LIFELINE. There's no TV, no telephone and the mail might come only four times a year. But there's always the marine band radio (antenna above). With it, the family can pick up messages all the way from Seattle.

it, latching it open.

"Quick!" I screamed. "Bring the buckets!" As soon as Randall dashed out, I pulled the inside door shut and raced after him.

The coal shed was on the leeward side of the house. Even with that protection, the wind clawed into us, scissoring the ties on our hooded sweatshirts up under our noses, whipping their plastic ends into our eyes and cheeks like bullets, yanking on the empty buckets like a demon.

Inside the shed, we filled each bucket with 25 pounds of coal. Then, laden with a bucket in each hand, we braved the wind. The weight helped, but when we came around the corner of the house, the wind struck us with all its force. Staggering and lurching, we fought our way back. I held the door while Randall lugged the buckets through it.

We rested. I retied our hood strings and jammed the ends under our coats where they wouldn't lash our faces. "That is some wind," I gasped. "It must be blowing 70 knots, at least."

"Yeah—and we still got to go to the barn," Randall wiped his running nose with the back of his gloved hand. "Have you seen the dogs?"

"No, but they'll be in their beds," I guessed. "All we'll have to do is lock the door." I peered out the window toward the barn and the two gates we had to get through. Off to the left was the chicken house with its heavy plank door. We would have to open the feed house, too, and get mash and milo for them. I shifted my gaze to the big wooden feeder in front of the granary. "No horses to feed either. They'll be hiding in a draw somewhere."

"What about our supper?" he asked. "Can we get some meat?"

"We can try," I told him. "The big warehouse door is in the lee. What do you want?"

"Chops," he grinned. "What else?"

I got the meat saw from its nail behind the stove and put a dish towel in my coat pocket to wrap the meat in and we set off once more. We walked abreast into the wind with our heads down and our feet braced wide apart, holding onto each other and combining our strength to open the gates.

At the chicken house, Randall forced the doors open and I

SCRATCHIN' OUT A LIVING. Fresh eggs are available as long as Cora keeps the hens fed and happy. Cora learned to be mother, wife, cook, baker, butcher, wrangler, shepherd, teacher and, eventually, even a spinner of yarn. There was little room on the island for anyone less self-sufficient.

fed the hens and gathered the eggs. A banshee gust caught me coming down the steps and I lost my balance and broke all six eggs, but caught myself before I fell on the saw I was carrying.

In the barn we located all four dogs and shut them in their pens. Since Milt kept a cull mutton carcass hanging for them to chew on, they didn't need feeding. Each one had his own space and went to it without coaxing.

"Good night, Floyd," Randall said. "Good night, Sambo. Good night, Sydney. Good night, Andy."

Next we followed the wide plank walkway to the wool warehouse, where our house meat hung from the rafters. Hacking cuts off a sheep carcass had seemed strange at first, but I had gotten used to it. I sawed double chops off the backbone while Randall held it steady without even thinking about it. Wrapping them in the dish towel, I stuck it under my coat.

Our most difficult chore was starting the generator. We had to retrace part of our steps to reach the generator house and go through one gate. The wind was at our backs now. All we could do was run with it, fetching up breathless against the fence, pulling with all our strength on the gate, then dashing through and digging in our heels on the other side so the wind wouldn't destroy the gate by smashing it against the posts.

The generator house lay behind the chicken house and next to the cattle chute on the round cutting corral. The walk was narrow and close and the only place on the ranch where I caught myself looking over my shoulder. It didn't give me goose bumps, yet I sometimes felt a presence there, something not quite in the same dimension as myself, as if someone from centuries past was watching. Chuck noticed it, too. Neither of us liked to walk that path after dark.

The generator itself was a 5-kilowatt Lister diesel, manufactured in England, a squat monster with a hand crank that took both my arms to lift. Besides turning the crank fast enough to start the engine, we had to flip two cylinder switches right when the engine reached the correct revolutions to fire. This meant taking one hand off the crank at the crucial moment, flipping the switches, and jerking the crank off the shaft before the engine really took off and turned the handle into a lethal weapon.

"**I'll do it,**" Randall offered, taking the crank from its shelf and fitting it over the shaft end.

"Okay." I squeezed past him to the front of the engine and pushed the two cylinder switches to their starting position, then stood with my fingers poised to flip them back the second I heard the engine catch.

After two bouts of wrestling with the crank and not quite making the correct engine pitch, Randall shucked out of his coat and gritted his teeth. "Get ready," he rasped. "I've got it warmed up now." But it was no use; he had used his best strength on the first two efforts.

"My turn," I said. We traded places. Unlike Randall, I knew I only had one chance. If I didn't get it up to the right revolutions the first time, I'd never have energy to do it again.

It wasn't having electricity for lights and a phonograph that was so important, although that was nice, but we needed the 4 hours of daily power to keep our freezer cold enough to preserve butter, cheese and bacon. More important, it kept our sideband radio batteries charged up so we always had communication with the outside world.

I stripped off my coat, spread my feet and grabbed the handle with both hands like a baseball bat. The trick, Milt had told me, was to gain momentum with every turn. "Mind set, Cora" I told myself. "Home run time."

I put my entire body behind the swing. One. When my hands were shoulder level again, I pushed harder. Two. Like music, I heard the rising crescendo. Down between my braced legs and back up again. Three. The engine screamed. "Now!" I shouted.

Randall pushed the switches while I swung the crank a fourth time and jerked it off the shaft. A roar filled our ears as the big motor surged to life. A small triumph, perhaps, but Randall and I whooped like we had swum the English Channel. We weren't just women and children—not us. We could take care of ourselves.

As soon as the engine leveled off, I threw the power switch for the house and outbuildings and we dashed back out into the wind.

POWER PLANT. Milt checks the oil in the diesel generator. Electricity, like most everything else on the island, has to be made. Cora and Randall had a time getting the old hand-cranked generator going, but it was either that or sit around in the dark. There are no electric companies in the Aleutians!

TOASTY TOOTSIES. Randall relaxes during a rare bit of time off from work and studies, letting the cookstove warm his feet and a good book fill his imagination.

The house seemed empty without Milt and Chuck as darkness approached. Although they were the quiet members of our family, the supper table wasn't the same without them. After the dishes were done, I offered to play a game of Yahtze with Randall.

"No," he shook his head. "I want to read." I thought he would bring out his ragged trapping manual, but instead he picked up *The Keep*, pulled his chair over to the stove and settled himself with his feet in the oven. In a few minutes, he was deep in the Transylvanian nightmare, emerging only to ask me now and then what a word meant or how it was pronounced.

Left to my own devices, I sat down in front of my spinning wheel, but somehow I couldn't dredge up any enthusiasm for an evening of frustration and failure. So I took out my drop spindle and a rolag of raw wool. The spindle was rough and homemade. In fact, Randall had made it for me from a wooden buoy, a stick of cedar kindling and a nail. I loved it.

With an expert flick, I set the whorl in motion and watched the twist run up the fibers of wool I held in my right hand. With this simple tool I could make yarn that rivaled a spider's web in fineness and strength. Why couldn't I do that on the wheel? Just because I had no one to teach me was no excuse. I had taught myself to cook out of a cookbook and I should be able to do the same with spinning.

Making yarn on the drop spindle was fun, but making it with the spinning wheel was fast. I wanted to spin lots of yarn, enough to make a rug. I already had the pattern picked out. And

WOOLEN WARMTH. Cora, with the drop spindle Randall made, has found a use for some of the wool they raise. She made the sweater she's wearing and the hooked rug from wool, and the braided rug from old army blankets.

I wanted to make sweaters and afghans. I wanted to use that mountain of wool I had been squirreling away in the loft of the barn.

Both Randall and I jumped when an onslaught of rain hit the window like a fusillade of bullets. "Uh oh. Who's going to turn off the lights?" He leaned down to pick up the book he had dropped.

I put down my wool. "Both of us," I said. "And right now, before it gets worse." I lit a coal oil lamp and headed for my coat. "I'm not going outside in this," Randall declared.

"Oh yes, you are," I told him as I pulled a heavy slicker over my head and picked up the flashlight. "Come on."

"You don't need me to pull one little switch; you're just afraid of the dark," he accused. But he got up and struggled into his wraps.

"I need you to help me get through the gate." That was only partially true. "And I am scared of the dark," I admitted. "None of us is perfect."

I motioned him to go through the door ahead of me. "Just be glad I don't make you do it by yourself."

"I'm surprised you don't," he grumbled. Then the wind and rain enveloped us with all its wet, lashing force. We struggled through it and returned to the house soaked and breathless.

I took the lamp and lighted our way out of the kitchen. Standing by the door to his room with the lamp, I waited for him to light his own. When it flared, I took the lamp on into my bedroom.

Later, after I was in bed, I saw the light under his door go out. "Good night, Randall," I called.

"Good night...Attila!" he thundered back.

LOFTY LOOMING. In the loft of the barn, where the wool is kept, is a good place for the loom used to turn that wool into useful items. Cora not only spins and looms, she dyes her fabrics using native plants.

Chapter Seven

A WINTER BABY

Without Milt in bed beside me, the wind's keening wail sounded close and lonesome. Even with Randall in the next room, the house seemed empty and vulnerable. When I was alone during storms, I often thought about the soldiers who'd been stationed at the base across the bay during World War II.

In their tents with only Sibley stoves and sleeping bags to keep them warm, those boys must have learned the real meaning of loneliness.

Later, when they had Quonset huts dug into the hillsides, the comforts still must have been scant. A few of the huts were still intact enough to tell what their construction had been like—only corrugated tin with inside walls of 1/4-inch Masonite. How the wind must have howled around them. How cold they must have been.

I pulled my warm woolen blankets and down comforter around me and closed my eyes. It was no use—I couldn't sleep with the noise. Somehow when Milt was home, the storms never seemed so loud.

I dressed and took my lamp to the kitchen. When Milt radioed at 5 a.m., I'd just finished scrubbing the porch floor and was taking cinnamon rolls out of the oven.

He sounded fresh and awake, but I suspected he'd been up all night, too, watching the boat. We didn't have any way to get it out of the water at the slaughterhouse and had to leave it tied to the dock. Any change in the wind had him out of bed and down at the water's edge. Sometimes I wondered why he didn't just sleep in it.

"Doesn't look like this is going to let up today," Milt said. "And I'm worried about the Ferret. We tied a tarp over the engine, but with this wind it's probably long gone. I don't like leaving it out in the weather and we might be stuck for days. I want you and Randall to go get it."

The request made me nervous for several reasons. The Ferret was a track machine. It had no steering wheel—only two levers. There were no roads to Cutter Point, just hills, tundra and bogs. The little machine was new. It had replaced our team and wagon 2 months earlier, when Milt decided it was time to move into the 20th century and retired the last elderly horses broken to harness. I'd driven it twice around the barnyard and never started it without it making a dreadful *errrrrrch* sound.

I took a deep breath. Well, I'd learned to drive a car; I could figure it out. "Sure, Milt, we'll get right on it."

"Ah…" There was a long pause before he answered in his most diplomatic voice. "Maybe it would be a good idea if Randall drove it; he knows how to start it. Just don't let him take it out of first gear."

"What you're saying is, I can't drive it, and Randall's not supposed to hot-rod?"

"Something like that," Milt hesitated again, then chuckled. "Sorry."

I often wondered why I could set up and monitor sophisticated medical machines with a great degree of skill, but when it came to a simple internal combustion engine I turned into a moron. And it wasn't just the engine, either. I had flunked my first driving test three times. Something in my brain just turned off when I got behind the wheel, and when that wheel was two sticks, it got worse. "Someday I might surprise you, Milt," I told him in all earnestness.

"Please don't do it today," he pleaded before signing off.

As soon as it was light enough to see our feet in front of us, Randall and I set out around the beach for Cutter Point.

I loved walking on the beach in a storm; there was something exhilarating about the breakers rearing up and rushing the shoreline like relentless armies, dashing themselves against the rocks in never-ending succession. I loved the spray on my face and the wind whipping and pulling at my clothes.

When we crossed Bishop Creek a half mile from the ranch, Randall tugged on my arm. "Look, Mom. A fox." There in front

FOXY VISITOR. *Although they're beautiful to look at, especially when their pelts are prime, foxes mean trouble for newborn lambs. They also can cause a ruckus in the henhouse if it's ever left unlocked.*

of us in the dim light, a red fox pawed in the piles of seaweed brought in by the storm. We were upwind and the pounding water deadened the sound of our approach. We walked to within 10 feet of him before he saw us. Like a streak, he whirled and bolted up the bank.

"Look at that coat!" Randall exclaimed. "It's prime."

I watched the bright plume of his tail disappear over the bank. Yes, he was a beautiful creature.

I knew they killed lambs, especially winter ones, often sneaking into the flock and grabbing the first twin while the ewe was giving birth to the second one. We tried to prevent winter lambs by keeping bucks out of the flocks from July through December. But with our sheep ranging over so many miles of tundra, we always missed a few bucks in the summer gathers.

Besides the havoc the foxes caused our sheep, I knew they stole chickens every time I forgot to lock the henhouse at night. I had watched one carry off our last goose in broad daylight.

The pelts, stretched and dried, brought a good price. Until a few years ago, the federal government had sent a professional trapper out each year whose sole purpose was to catch foxes. Besides the losses we sustained from them, they also ate gull eggs and any nestlings they happened across.

The fox appeared again high on Cutter Point hill, a blur against the brown grass, his tail whipping in the wind. Part of me said, "Run away, little fox, this is a dangerous place," while another part of me asked, "How many of our lambs and chickens nourished that gorgeous coat?"

On around the coastline we hurried, hardly noticing the wavering lights of the processors through the smoking spray, so familiar had they become. When we reached the Ferret, the tarp was indeed gone, but Milt had parked it behind the Cutter Point warehouse so the machine was somewhat protected.

I know the Bible says we are all created equal, but when it comes to machinery I do believe that males have the edge. (At least the males in my family have the edge over me.) Randall started that engine on the first try. He leaped into the driver's seat and motioned for me to get in the passenger seat behind him.

The machine was built with wide running boards over the tracks at about waist level. Between them were the two seats, one behind the other. In front of the driver's seat were the two levers for steering; pull back on one and it would turn left, pull back on the other and it would go right, pull back on both and it would stop. Stuck high on the front was a 15-horsepower Kohler engine. In first gear it went about 4 miles an hour.

"Don't you try to scare me," I hollered, clambering on but not in the seat. I perched on one running board with my back to Randall in a position that best afforded instant evacuation if I deemed it necessary to abandon ship. There was no cab; it was open all around, a feature I liked.

Off we lumbered up the steep hill behind the warehouse. The machine crawled over the uneven ground like a tank. Dispersing its bulk evenly over wide tracks, it waded right through Cutter Point bog, where no one dared take a horse. That was another nice feature.

On the minus side, it bucked worse than a wild horse as it climbed over the high tussocks of frozen grass. I braced my feet on the hitch and hung onto the rail with both hands. When we stopped without warning, I cannoned into Randall's back.

"What's wrong?" I yelled.

He pointed above us to the left. Through the blowing rain I glimpsed a small band of sheep huddled on the hillside. A quick flash of color caught my attention; just under the summit was the fox we had seen earlier.

One sheep lifted its head and they all bounded away, around the hillside and out of sight. There was a straggler a short distance below the main bunch. She hesitated for a second, watching us, then with a plaintive bleat, she followed the others.

The fox ran, too, but not far. Something dangled from his mouth. A sudden premonition seized me. "Let's go see what's up there." I jumped off the Ferret and ran, tripping on my raingear, falling against the sharp incline as I started up the hill. I struggled to my feet and Randall passed me.

This is stupid, I thought, as I pushed myself away from the hill. That fox had a leg in his mouth; the damage is already done.

Above me I heard Randall shout. When I reached him, he was kneeling in the grass beside a shivering lamb, so newborn its

FOLLOW THE LEADER. Sheep could be just about anywhere on the ranch, and when it came time to herd them in, Cora, Milt and the boys got into all kinds of adventures. Rugged and rocky terrain like this often made the job that much tougher.

fleece was still bright orange. A few feet away were the remains of its twin.

Randall hadn't touched the lamb; we knew better than to leave our scent on him in case his mother came back. If she smelled us, she would never claim her baby. I scanned the hillside for the ewe, but she was nowhere in sight.

Just then the lamb gave a wavering "maa-aa" and struggled to its feet. It was a sturdy buck lamb, but his sides were all caved in, so I knew he had never sucked. Without that first colostrum, he would soon die.

What should we do? If we left, the mother might come back. But how long would that take? The fox was still close by. He, too, would return the minute we were out of sight. It was raining, and with the north wind, the chill factor was below freezing. The lamb would die soon just from the cold.

But if I took him, the ewe stood a good chance of getting mastitis from a tight bag; then she would be ruined for raising more lambs. What would Milt do?

I motioned Randall to shield the little body from the wind and I hiked to the top of the hill. The sheep were gone; the ewe with them. She had probably been so frightened by the fox she would have abandoned her lamb anyway, I told myself.

My decision made, I slid back down the hill. There had never really been any question. I had known from the moment I saw the tiny animal I would never leave it to the vagaries of the wilderness. Randall had known it, too. When I got down the hill, he already had him wrapped up in his sweatshirt.

I put him inside my rain jacket with his nose against my neck and walked the rest of the way home. The lamb didn't struggle; didn't even move. I couldn't feel him breathing but dared not unwrap him again just to see. I jiggled him up and down in my arms to stimulate his respiratory muscles.

When I got to the house, Randall was already there and had a box all warmed up on the oven door. I pushed the bundle into his arms and shucked out of my raingear, leaving it in a sodden dripping heap on the porch floor.

"I think he's dead," Randall called from the kitchen.

Poor thing. He certainly looked dead. His legs flopped out in all directions, his head lolled back with his eyes open and staring. I put my hand on his chest. Under my fingers a faint pulse still fluttered, but he wasn't breathing, and he felt so cold.

I pushed down on his ribs a couple times and he made a feeble gasping sound. "Only mostly dead," I joked, trying to reassure Randall. "Fill the sink with hot water." He rushed to the tap. "But not hot enough to burn your hand," I added.

I lowered the limp body into the sink until all but the head was immersed. Gently I continued squeezing the frail chest without any reaction until I despaired ever saving him. If warm water and artificial respiration didn't work, nothing would.

All of a sudden, mucous bubbled from his nostrils as the air rushed out of his lungs. He emitted a strangled cry and drew in a good breath. His legs flailed and splashed us both with water.

"He's alive!" Randall cried.

"Get a towel," I told him.

When he returned with it between his hands, I lifted the lamb into it. "Now rub his legs," I instructed. Together we massaged his limbs and torso until he was dry, then Randall laid him in the box.

"Look, Mom. He's sucking on my fingers."

"Great, he's hungry," I said. "He needs milk." I diluted an ounce of evaporated milk with an ounce of warm water and dribbled some into his mouth. After a couple attempts, he grabbed the nipple and pulled on it eagerly. He had a strong suck, and when the warm milk hit his stomach, his tail started wagging.

"He likes it," Randall whispered. "I want to do it." I handed him the bottle. "Do you think Milt will let us keep him?" He touched the tight matted curls on the lamb's head.

"I don't know," I answered. I had been wondering the same thing. "We'll just have to ask him."

In truth, I doubted he would be pleased at my actions. Bum lambs were a big nuisance, he had told me the first time I wanted to steal one from its mother, and they never did as well as their wild counterparts.

But this was different, I argued, unconsciously marshaling my defense; this time the mother ran away. I couldn't just let the poor thing die.

As if reading my mind, Milt called on the radio. First I told him we had gotten the Ferret home safely, then I sprang the news about the lamb. I could tell by the lengthy pause that the news was less than welcome.

"You're sure the old ewe wasn't coming back?" he asked.

"And there was a fox, too," I added. "He ate the twin."

Randall yelled from the doorway, "Ask him."

BOTTLE BABY. There weren't many things more cute than "bum" (or orphaned) lambs and Cora just couldn't leave them alone to fend for themselves.

I took a deep breath. "May we keep him for a pet?"

"We'll see," he stalled. "I suppose you're feeding him canned milk."

"We just tried him on a little," I explained eagerly. "He likes it."

"Well, don't get your hopes up," Milt cautioned. "He'll probably run up a big grocery bill and then die." Milt's warning came too late. Our hopes were already sky high.

There is nothing in the world cuter or more affectionate than a baby lamb. Randall and I lost our hearts to that tiny living scrap. We became slaves to his every whimper and bleat. Between classes, we fed him and changed his bedding.

"What shall we name him?" Randall asked.

"Let's just call him 'Lamb' for now," I suggested. "He'll name himself in a few days."

By evening, he could stand by himself, and the next morning he made an attempt to gambol after Randall took him out of the box. But the milk must have been too rich, because it gave him the scours. I cut back to a couple tablespoons every 2 hours. He gobbled it up and wanted more.

We were so excited we could hardly wait for Milt and Chuck

to get home so we could show them. But the storm raged on and it was the evening of the third day before they were able to travel across the bay. By then the lamb had made himself at home in the kitchen and attached himself to me. As far as he was concerned, I was his mother. He stayed beside me whenever he was out of his box.

When Milt and Chuck came through the door, tired and hungry, the lamb rose to the occasion with a magnificence far beyond my expectations. As if he knew somehow that he was on trial, he left my side and greeted the newcomers with a friendly "maa-aa", a sideways jump and a series of stiff-legged hops that sent him careening into the cupboards.

"Hey!" Chuck exclaimed. "He looks pretty lively to me." Then he sniffed. "Pee-yuu."

Behind him Milt groaned. "Get used to it, Chuck. Mama has a new baby." But he knelt and patted the lamb's head. "Come here, fella." He ran his hand over the fleecy white coat. "Nice big lamb," he said. "Too bad you didn't get there in time to save them both."

While we ate supper, Randall told Milt and Chuck how we had rescued the lamb. "You should have seen that fox, Chuck," he enthused. "I can't wait to start trapping." He attacked his cherry pie. "Just as soon as my traps come, I'm gonna catch that fox."

"Not with those traps you ordered," Chuck snickered. "You aren't strong enough to set them."

"Am too!" Randall shot back. "And I'm putting my first set right by the creek."

WHAT'S FOR SUPPER? While Milt "feeds" the cookstove, a pet lamb waits impatiently to be fed. Orphaned lambs, starting with Peep-Sheep, often became part of the family.

"Who says?" Chuck's voice took on an ominous tone. "You don't own the creek."

"Yeah, well neither do you," Randall retorted. "I'll put my traps wherever I want to."

"That's enough," I said firmly. "No arguing at the table."

The boys subsided but continued to look daggers at each other as they finished their dessert. I had worried about this happening. Now that the season was here, I knew something would have to be settled or they would be quarreling over sites all winter.

I got up and cleared the table. The lamb followed me back and forth to the sink, staying as close to my knee as he could get. Even when I only went a few steps, he was still right there.

"This little guy thinks he's a person," I said, sidestepping so I wouldn't run over him.

"Yes, he's a people sheep now," Milt grumbled. "He'll never be anything else." He snapped his fingers. "Come here, People-Sheep, get out of Mama's way." Scooping up the lamb, he carried him to his box. "What do you think, Mom? Shall we call him 'Peep' for short?"

"I like it," Randall chimed in.

"Sounds good to me," Chuck agreed.

Peep-Sheep it was.

Chapter Eight

A BARN FOR ALL SEASONS

When I woke up, the storm was over and bright sunlight spilled in through the windows. I turned to see the clock and Milt didn't even stir. My gosh, 10 already. The lamb must be starving. I slipped out of bed.

A warm blast of air hit me as I opened the kitchen door. Coffee simmered on the back of the stove. Chuck looked up from setting the table and Randall pulled the empty nipple away from Peep. "Good morning," Chuck said.

"I thought you were going to sleep all day," Randall grumbled.

"Thanks," I waved my hand to include fire, coffee, table and lamb.

Much later, as we finished breakfast, I asked Milt about the boat crossing. "How bad was it?"

"About as rough as I ever want to be out in," he admitted.

"I thought we were goners when I saw that log," Chuck declared.

"A miss is as good as a mile," Milt said with a warning look in Chuck's direction. He pushed back his chair. "I need some help with the sheep." Both boys followed him out the door. I knew I had heard as much as I was going to about their boat adventure from Milt.

After I finished the dishes, I fed Peep again and took him with me to the barn. As I maneuvered my small charge through the gate from the shearing floor, I heard the dogs barking and Milt's low-voiced commands.

From the gate, I watched the boys pushing the flock from the high drying floor in the back of the barn to the lower floor close to the holding pens. I liked working with the sheep. We raised Columbias, for both meat and wool, but mostly for wool. They were a breed developed in Idaho around 1912 by crossing

Lincoln with American rambouillet. We liked their hardy constitutions and good mothering instincts, open faces that meant no blindness, and big frames that carried a lot of flesh.

Their wool graded medium, the finest we could raise in our harsh, damp climate. Any finer wool, such as pure rambouillet or merino, would never get dry enough to shear in our short wet summers. The Columbia fleece was white and soft, grew about 5 inches a year and averaged 8 to 10 pounds on a total diet of range grass and seaweed.

They ran the animals into a holding pen where they couldn't get away and then grabbed the ones Milt selected. "How about this one?" Randall gasped, his arms around the neck of a woolly mammoth with big curling horns. "I want his antlers."

"Antlers?" Chuck teased.

"Well, whatever," Randall shrugged. "I want them."

"I don't like antlers either," Milt grinned. "They're murder when you're shearing; put him in the alley."

CORA HAD A LITTLE LAMB. A bum lamb's fleece might not always be white as snow, but everywhere that Cora went, the lamb was sure to go, as this one is doing.

In the next half hour, Milt chose eight animals and let the others go back to Pasture 5. The eight he selected we put in a smaller pasture beside the barn.

Just as Milt had predicted, Peep showed no interest in his own kind, staying glued to my side the whole time. I offered to close the gate on Pasture 5. I walked up the hill in the sunshine with Peep bounding along at my knee. As I struggled with the wire loop fasteners, Randall caught up with me. "Can I play in the barn with Peep for a while?" he asked, adding his weight to mine on the gate.

"Yes," I said. "But we must do lessons this afternoon." He dashed away with Peep in his arms, promises to study floating back over his shoulder.

After being cooped up in the house for 3 days, he needed a break, and the big barn was a wonderful place for exploring. **I liked it, too.** The front shop, full of old tools, fascinated me. I could spend hours just trying to identify the implements hanging from the walls and rafters. Besides the saws, wrenches and hammers, a hand drill with cogs big as my fingers stood bolted to the wall beside a cast-iron forge with a hand-driven fan and a grinder whose power came from a pedal. These were things I didn't know existed until I came to the ranch, but things Milt used every day.

Amid all those tools, the place of honor went to a huge blacksmith anvil mounted on a driftwood stump. It had holes drilled along one side and strange pointed hammers fit into them, for hammering on horseshoes, Milt told me. I just liked to clang them against the iron anvil and listen to the music, even though he told me banging on a cold anvil with a hammer was not good for the metal.

In fact, he informed me further, there was an old saying among blacksmiths that emphasized the point: "A good smithy never hammers on cold steel or he doesn't go to Heaven." Still, I liked the sound and couldn't resist a ping or two in passing.

Under the stairs to the loft was the tack room, with each tree holding a worn saddle. It smelled of leather and horses and freedom. The rows of bridles and halters jingled as I brushed against them.

Seven stalls with deep mangers and sculled plank floors filled the long room opposite the shearing floor, conjuring up soft whinnies and stamping hooves even when the stalls were empty.

The main area was called the shearing floor and ran about a quarter of the length of the entire barn. There was room for four shearers; each station contained an electric shearing drop with motor and jointed hanger. Suspended from the rafters above each station was a circular hoop called a sling. It supported the shearer's weight during the long hours of bending over the sheep.

When I first learned to shear, I tried to use the sling, but the tension was too strong. My feet came off the floor and I zoomed around my station like Peter Pan until Milt noticed my predicament and rescued me. My sheep, of course, escaped entirely.

The shearing stations were sandwiched between a long alley, with holding pens for unsheared animals on one side and small enclosures with openings to the outside for the sheared sheep along the other. Behind it were the crowding pens and drying floors where we had separated the sheep.

Instead of going back to the house, I climbed the ladder to the loft. This was my special place in the barn. It had a low pitched roof that made me stoop to walk except in the very middle. Rows of fleece lined the walls, stacked on wire trays, piled on the floor and hanging from the ceiling. In one corner by the ladder Milt had a partitioned area for all his pack gear, saddles, slings and ropes.

There was a tiny window in the front where I could look out over the barnyard and bay. Fiberglass panels let light and heat in through the metal roof. It was a warm, quiet place where I could be by myself if I needed to. And there were times I needed just that; we all did, and the barn was big enough for each of us to have room to be alone and still together.

High and private in my eagle's nest loft, I could listen to Milt working in his shop below, hear the dogs barking and playing in their little rooms scattered throughout the building and watch the boys shoot baskets in the hoop above the lower drying floor.

I pulled out the fleece I was working on and ran my hands

LOFTY SORTING. *Wearing a sweater spun from Chernofski wool, Cora sorts some more raw material in the loft of the shearing barn. Cora eventually got the hang of shearing and spinning.*

deep into its lanolin-rich fibers. Snow white and a very fine grade for our flock, I wanted to make something wonderful from it—a sweater with big sleeves and a hood—if I could just learn to spin on that wheel. It would take me forever to make enough yarn on the drop spindle. I sighed, knowing I should be in the house practicing on the wheel instead of out here daydreaming in the barn.

Wandering to the window where the sun was streaming in, I rested my elbows on the sill and looked out. The water had calmed down until there was only a small surge foaming along the beach. As I watched, two fishing boats appeared around the headland and wallowed through the channel. There was still a big sea running, but I knew the fishermen were anxious to get back on their gear. Time was money to them, and where king crab was concerned, big money.

I saw Milt below me on the plank walkway between the barn and yard gate and went to meet him. "Ah, there you are," he said. "I need your help with the grease trap."

"The grease trap?" I fell into step beside him. "Why are you cleaning that today? I thought you were going to take a nap."

Cleaning the grease trap was one of those jobs that even Milt put off as long as he could. It was a 30-gallon barrel buried beside the house with the kitchen drainpipe running through it. All the grease and fat from the sink collected in it instead of running through the main drainpipes to the septic tank. It was a smart system and saved a lot of plumbing headaches and digging up pipes, but it sure was a messy job to clean.

The theory behind it was that fat rose to the top of the water and the pipe ran through the middle of the barrel. We skimmed the congealed fat off the top with a deep ladle Milt had made for that purpose. There would be buckets of it to be carried into the horse pasture and dumped on the ground. The seagulls and ravens loved it.

After skimming the fat, we had to clean out the drainpipe between the trap and the main connection to the drain field. To do that, Milt left a foot-long spring tied between two 30-foot lengths of rope in the drainpipe all the time. That is why he needed me, to pull on one end of the rope.

"Well, why aren't you taking a nap?" I persisted as he opened the gate.

He squinted his red-rimmed eyes and looked at the cloudless blue sky. "Because I feel a snowstorm coming and I want this done." We walked around the front of the house and I saw he had already cleaned out all the grease.

"You should have called me sooner," I bristled.

"Chuck dumped the buckets," Milt said, handing me a broken shovel handle looped around one end of the rope. "He's tired, too. I told him to take a break."

"Well, all right," I said, taking the handle and wrapping a loop of rope around my waist for better purchase. Chuck wasn't afraid of hard work, but he liked excitement. When chores like grease trap cleaning came up, he sometimes dropped out of sight entirely.

Milt disappeared around the corner of the house. In a few seconds I felt a tug on the rope. "Pull!" he yelled.

I walked backward, exerting pressure on the rope until I heard the spring falling into the grease trap water. "Okay!" I shouted.

Back and forth we pulled the rope, yelling at each other, stopping short when the spring caught up on a corner of the pipe, jerking and tugging until it came free. We were so engrossed we didn't hear the outboard motor or the footsteps coming down the walk.

"Hello."

LOVE IS SHARING KP. Chores at Chernofski range from the exciting to the mundane. The Holmes family knows that sharing even the simplest of them, such as the breakfast dishes, makes them a lot easier. Besides, it's a lot cleaner than bagging coal.

I looked up in surprise. Two unfamiliar young men stood just inside the gate. Both carried large green bags Santa Claus-fashion on their back. The mail!

"Merry Christmas!" they called.

I gave the rope two sharp tugs and hollered for Milt. Then I dropped it and met the men halfway as they crossed the yard with outstretched hands.

"Hi, I'm Cora," I said, grasping the nearest hand.

The dark-haired fellow pumped it vigorously. "Boy, have I heard a lot about you." He continued to shake my hand. "I'm Jim Severns, ma'am, superintendent on the *Akutan*, and this is Kim." He nodded toward the other man who looked about 18 years old. "We've really been looking forward to meeting you." Both flashed me wide grins, their faces full of admiration.

Bewildered, I gave them an uncertain smile in return, wondering what in the world they were so excited about. Suddenly it dawned on me, and I burst out laughing. As Milt came around the corner of the house, I called to him. "Milt, come tell these guys that I don't butcher all by myself."

"Hi, Jim. Kim." He shook hands with the two men, obviously already well-acquainted. "Sorry to disappoint you." He threw an arm across my shoulders. "But she does make great cinnamon rolls. Come, have one."

Inside, Chuck poured coffee while I got out the rolls. "Thank you for bringing the mail," I said. "We wondered who had it."

"We had a plane the day before the big storm hit," Jim explained. "But this is the first chance we've had to bring it over." He bit into his roll. "Sorry."

"Don't be," Milt said. "We really appreciate your getting it." He shrugged. "We're used to Christmas anytime between January and March."

Randall burst into the kitchen, Peep on his heels. Stopping short when he saw the two men, his face fell. "I thought the girls were here to ride George," he said.

"Not today," Jim said, "we don't have time." He pointed to the sacks and Randall brightened. "But we'll bring them over soon," he promised. "For one thing, we might not be here much longer."

"The *Akutan's* pulling out?" Milt asked in surprise. "Why?"

"Not as much crab as usual," Kim answered. "The catch is way down from last year."

This was unexpected news. For the past 10 years, fishermen had been crabbing in Umnak Pass and the surrounding waters with incredible success. Even small boats sometimes topped a million pounds a season. Everyone thought the supply was inexhaustible. I looked at the men's faces, suddenly turned serious. Crab was their livelihood, too, even if they didn't actually catch them.

"Maybe it's just an off year," I suggested.

"I hope so," Jim said, draining his cup. "We all hope so."

After they left, we cleared the table and emptied both mail sacks out on the floor. "It sure will curtail our mail service if the boats all leave," I remarked to no one in particular, folding the empty sacks and setting them to one side. "But, on the other hand, it will be nice and quiet around here again."

Milt looked up from the jumbled pile of catalogs, magazines and parcels. "They've been fishing the Bering Sea hot and heavy for a long time now. It's bound to have diminished the breeding stock." He glanced across the table at the boy's solemn expressions.

"Will it make a difference in the crab pot business?" Chuck asked.

"Not this year," Milt answered. "But yes, if the crabs leave, there won't be any reason for crab pots."

"Oh, no," Randall groaned. "Who will buy my fox hides?"

"We'll still have boats the rest of this winter," Milt reassured him.

"Then I've got to start right now," he stated. "Where are my traps?" He swiveled his head toward me. "Come on, Mom, hurry up!"

"Not so fast." I handed him a thick packet of magazines and letters all fastened together with rubber bands. "You can help sort mail first." I slid a similar package across the table to Chuck, and for several minutes everybody talked and nobody listened.

Sorting mail had become a familiar task in the last 2 years. Everyone knew the routine: magazines in one pile, catalogs in another, junk mail on the floor, packages on one end of the table, personal and business letters on the other.

When all the banded packets were sorted, I turned to the packages while Milt started in on the pile of business letters. Chuck stayed in his chair at the table, but Randall had no such reservations. He knelt on the floor beside me and reached for the first box, which had the distinctive red label of the correspondence program.

"That's Chuck's," I said, and he pushed the carton toward his brother with a fiendish chuckle.

"Now you have to go to school again, same as me."

"So, who cares?" Chuck replied, looking up from the letter he was reading. "It's no big deal." But he didn't make any effort to open the box. He didn't even look at it. "Is that all I got?"

"Nope," I grinned, handing Randall an oblong cardboard package with his name on it before tossing Chuck a big carton from L.L. Bean. "I wonder what this could be?" He caught it in mid-air and shook it.

"At last," he beamed, "my boots." He ripped the box open

with as much enthusiasm as Randall, pulled out the heavy pacs and tried them on.

"Oh, boy, my traps!" Randall shouted, tearing the box top off and dumping out the contents. "Look at these, Chuck."

Everyone occupied, I went back to the letters, sorting out the ones I wanted to read first and giving the boys theirs. I dropped Randall's on the floor beside him, noticing only a small envelope from the school, surprised that none of his corrected lessons had returned. He would be disappointed. I went back to my own letters, sorting them by postmarks.

There were three from my mother and one from my sister Doris. Mom had sent a large Christmas package, which I knew would be filled with all the love and good cheer she and the rest of my family living in Idaho could stuff into it.

I opened the earliest envelope and read the cheerful, news-packed pages. I came from a large family, the seventh of 12 children, seven girls and five boys. As we grew up and went our own ways, we didn't keep up on each other very well, especially me, living so far away from everyone else.

But Mom, a widow, kept track of us all, and through her "Gorley Grapevine", she relayed any news that came her way. I knew anything I wrote to her would soon circulate among my siblings. Now I read with relish all the everyday happenings of people I loved and hadn't seen in 2 years.

"Read it out loud, honey." Milt's voice brought me back to the warm cluttered kitchen from 2,500 miles away. He had pushed aside the business letters and propped his elbows on the table.

"Yeah," Randall looked up from his school letter and scooted closer. "I want to hear Grandma Gorley's letter."

"Me, too," Chuck closed his new Cabela's catalog.

So I read to them about their cousin Christy Jo going to Switzerland as an exchange student; about my little sister Ginger expecting another baby; about my brother Joe building a barn on his new property in Cascade, Idaho; and about the big Thanksgiving dinner Mom had fixed for 18 people in November, and how much fun they'd had selecting things to go in our Christmas box.

When I finished, Randall said, "I love my Grandma Gorley."

Chuck, usually reticent about his feelings, spoke up, too. "She is the perfect grandmother." He smiled sheepishly, "You know, white-haired, kind of round, always smiling…"

"And baking cookies," Randall interrupted. "Peanut butter ones with chocolate chips. Can we open her box now?"

"In a minute," I

MAIL, AT LAST. With mail delivery only three or four times a year, its arrival is quite an occasion. Cora relaxes with a letter from home, while the boys tear open boxes containing their orders from catalogs.

said, slipping the letter from Doris into the hip pocket of my blue jeans. "First I want to hear Milt's letter from Mildred." Although Milt's family wasn't as large as mine, only five people, he had a twin sister. A special bond kept them close, even though she lived in California and Milt hadn't seen her for years

Milt read about his brother, Tom, who had stayed on the family ranch in Midvale, Idaho, and how successful his feed bunk business was becoming. On an encouraging note, he read that his older sister in Texas was recovering from lung cancer surgery with a good prognosis; not so promising, his brother-in-law in northern California was being advised to consider dialysis.

The letter ended with the good news that both Mildred and her husband, Tony, were fine, and looking forward to the wedding of their youngest son.

I planned to read the latest letter from Mom, dated only a week earlier, but the boys expectant faces changed my mind. I put it in my other hip pocket for later.

"Okay, let's open Grandma's box." It must have weighed 30 pounds. Randall could barely heave it up on the table. Inside all

CHRISTMAS PAST. Chuck was 16, Randall 12, and Milt beardless when this photo was taken. The Christmas tree was a gift from a passing freighter. The decor was still "spartan bachelor" before Cora came down with what she calls "Frontier Victorian fever" and redecorated.

the packages were wrapped in bright Christmas paper. After Milt played Santa, all our favorite things came spilling out of the festive wraps. New slippers, caps, knee stockings, all hand-knitted in Grandma's intricate patterns.

Motorcycle and *Mad* magazines from Aunt Doris tumbled from Chuck's package; *Garfield* and *Archie* comics from Randall's. Soon the floor was covered with jigsaw puzzles, latch-hook rug kits, games, magazines, paperback best-sellers and knitting patterns.

Stuffed in all the crannies were small lengths of fabric, rolls of lace and rickrack, packets of needles, spools of thread and embroidery floss.

For Milt there were tins of root beer-flavored hard candy, lemon drops, a hard-to-find book by Ardis Fisher (his favorite Idaho writer) and the new Far Side desk calendar.

And from the bottom of the box came 4 pounds of Grandma's homemade fudge and a double recipe of her steamed Christmas pudding. It had been in the mail since Thanksgiving and had patches of mold clinging to its sides. My eyes filled; for a brief moment I saw my roly-poly mother taking the steamed pudding out of coffee cans and lining them up on her kitchen counter.

I loved my new husband, was happy in this wild isolated harbor and never wanted anything to change. But sometimes I did miss my family, especially on mail days, when they materialized so poignantly out of letters and boxes. Giving my eyes a furtive wipe, I glanced up to find Milt watching me.

"It's hot in here," I gave him a watery smile. "I think I'll take a walk."

"Want me to come?" He laid down the letter he had been reading.

I saw it was the one from his twin. He, too, had his nostalgic moments, I knew, when he worried about perhaps never seeing his brothers and sisters again, especially now that Peg was so sick. But he handled it better than I did, always maintaining that whatever happened was for the best, that to fret about things beyond our control was a useless waste of time.

I shook my head. "I won't be long." Grabbing my jacket, I escaped to the barn. Retracing my steps to the loft, I sat on a pile of fleece and opened my letter from Doris. Three years older, she had been from childhood my dearest friend and closest confidante; her letters always filled me with warmth.

I read down the closely written page with increasing disbelief. They had just returned from Wyoming, where they had attended a funeral. Basil, my sister Alice's husband, was dead. I closed my eyes against the words. He was only 37 years old. Alice was 33. They had three children and the youngest was 9. Why had this happened?

I ripped open the other two letters from Mom. It was all there. Basil, ill with a kidney problem, had contracted a blood infection and died after 3 weeks on a respirator in Salt Lake City.

Three weeks! He had been sick enough to die for 3 weeks and I didn't even know until after he was gone. Dry-eyed, I stared at the words of my mother's letter, numb and heartsick,

WHAT A VIEW! Mount Aspid, snow-covered from October to July, is the view from the kitchen window of the ranch house. In the harbor is the Maud, a small boat owned by the ranch in the early days. Penned sheep are ready for the dipping vat.

yet knowing that my presence would not have made a whit of difference. Still, I wished I had known, wished I could have added my support to that of everyone else who had gone.

What would Alice do now? I pictured her sitting alone in that 100-year-old log house on a ranch in southern Wyoming that was almost as remote as ours. I started wording a letter, saying the things I wanted to tell her in person, feeling helpless in the face of her loss, and somehow guilty because all the people I loved most were healthy and alive, and even now making noise in the shop below me.

I peered over the loft railing. All three of them looked up. Randall's face was tear-stained; the new traps dangled from Chuck's hand. "I'm up here," I called. "What happened?"

Randall held up his hand, still sniffling and wiping his eyes. Milt explained, "He caught his finger in one of the new traps."

"I told him they were too strong when he ordered them," Chuck added, swinging the traps against his leg so they jangled. "Now he believes me."

"Why do you have them?" I asked.

"I sold them," Randall hiccuped. "Milt said he had some I could use."

Milt looked up at me. "The government trapper left his; the springs are old and weak. All they need is a little fixing." He walked across the shop until he was directly under me. "Are you okay?" he asked.

I hesitated, then nodded, surprised that the numbness was wearing off. I would wait until later to tell him. "I'm fine," I said and, watching them rummage through a box of rusty parts, realized I had told him the truth.

I returned to my nest in the wool and gathered up the scattered letters. From below I heard the sharp metallic music of hammer hitting anvil. I lifted my head and listened. Yes, I really was okay.

IT'S A PAIR O' DOCKS. Actually it's two views of the slaughterhouse dock. The aerial view shows some of the many buildings abandoned when the military left the island after World War II. Cutter Point is on the other side of the bay and is the shortest distance across, which is handy when the water is rough. Ranch headquarters is about 1-1/2 miles from Cutter Point.

Chapter Nine

MILTON'S MAGIC

When we came out of the barn it was snowing. The wind had dropped to nothing and the flakes floated down around us in lazy spirals, covering the ground and making the quiet afternoon even more silent. The second-hand traps dangling from Randall's hand chimed like sleigh bells, muted and far away.

I told them about Basil as we walked to the house. Milt and I hadn't known him well, and neither Randall nor Chuck had ever met him. Still, the boys' faces became solemn and they didn't speak until we entered the kitchen and saw the magazines and crumpled papers strewn around the room.

"I'll go start the light plant, Mom," Chuck offered. "It's dark enough, isn't it?"

I nodded and he slipped back out the door. Moments later, the generator hummed to life and light flooded the rooms. Even though daylight had been gaining about 4 minutes every day since Dec. 21, it was still dark in the house by 6 p.m.

While Milt added coal to the fire, I pushed up my sleeves and waded into the mess.

"Cleanup time!" I announced. "Everyone picks up his own mail and at least four catalogs."

Behind me the door slammed shut and Chuck stomped the snow off his new boots with a resounding thump. When I turned around, he held up one foot, his face beaming with pride.

"Can I keep them on awhile? They're still clean."

"Aren't they heavy?" I asked, eyeing the cumbersome footwear.

"Nah," he said, shaking his head. "And my feet are toasty-warm. I might never take them off."

Randall tugged at my arm. "Mom, what does de...de...linquent mean?" He held out a crumpled sheet of paper.

I took the paper. "It means being behind in your bills, usually," I explained, glancing at the letterhead. It was from the school. As I read the terse lines, I became more and more perplexed.

"Your teacher says he hasn't received any lessons from you since school started in September and he is placing you on the delinquent list." I was talking more to myself than to him.

"What?" Randall shouted. "You write and tell him I already sent in one whole quarter! He must have lost it!"

His eyes turned desperate. "I'm not doing it over! You can't make me!"

He stared mutinously at me for a few seconds, then his shoulders sagged and his lower lip trembled. "Please write and tell him I already sent it," he begged.

I folded the letter and laid it on the table. "What could have happened to it?" I looked across the table at Milt, who was sorting through the business mail. "Could the post office have mislaid it, do you suppose?"

Milt tapped an envelope against the table before handing

HORSES, OF COURSE. In the winter, the horses are kept corralled near the ranch headquarters and allowed to graze only on pastures that were closed off in the summer. Saddle horses are also fed oats twice a day in the winter to supplement their diets.

it to me. "Maybe. But more than just the school stuff got lost. Read this."

It was a letter from our health insurance company, saying that our policy had lapsed because we hadn't paid the premium.

"I'm sure I paid that last fall," I told Milt. I hurried to the desk and got out the checkbook. Yes, it was there, paid in September, along with several other bills. "Did you get your generator parts?"

"No," Milt replied. "And there isn't a letter from them saying they're back-ordered."

"None of my Christmas orders came," Chuck broke in. "And I sent them out with the last mail before Thanksgiving."

I turned to Milt. "What shall we do?"

"Try to trace it, I guess." Milt rubbed his jaw. "Do you remember who took it out for us?"

"I do," Chuck said. "It was the *Man of War*, and they were going to Kodiak."

"Oh, no," I moaned. "We'll never find them. They probably tied up the boat for the holidays and got a whole new crew since then..."

"Hello all mariners! Hello all mariners! This is WBH29 with the national marine weather," Peggy Dyson's voice broke into our conversation from the marine radio.

Milt and I looked at each other and shouted in unison, "Peggy!" We dashed to the radio room.

For as long as I had been at the ranch, Peggy Dyson came on the air twice a day from Kodiak, Alaska with a marine weather forecast. And after she finished her report, she patiently repeated any areas that fishermen called in and requested.

Peggy also made phone calls and passed on messages for the men out at sea. And although Kodiak was 400 miles from us, and we had never met her in person, Peggy called us on Christmas and wished us a happy holiday.

Peggy knew everyone within a 2,000-mile radius of Kodiak, the names of their boats, their wives and most of their kids. If anyone could find our mail, Peggy could!

After the last weather report was given, Milt called Peggy and explained our problem. "If you locate the boat, ask them to look in any empty bunks for a green mail sack from Chernofski." She promised to do what she could.

For 2 days we waited for her call. Randall plodded through his studies with a glum face, sure all his work was lost forever. As often as every half hour, he would declare to the room in general, "I won't do it over...they can't make me."

Chuck, who had reluctantly started his second-semester studies, didn't believe the mail would be found and delighted in teasing Randall about all the makeup work he'd have to do.

But I had faith in Peggy. Any woman who had the patience to repeat weather warnings six or seven times after the regular forecast for fishermen who had forgotten to listen, was capable of finding one lost bag of mail.

On the third morning, my trust was rewarded. Peggy broadcast our call sign after her regular forecast and said she had located the boat. Our mail was now safely in the post office.

With a great shout of glee, Randall zoomed around the radio room like an airplane.

The snow continued to fall all day, and Randall and Chuck wanted to be out on the snow-covered hills that beckoned outside the kitchen windows. They even stopped bickering long enough to finish their lessons early so they could go sledding before dark.

With binoculars, I watched them head for Cutter Point Hill, pulling a 4-foot square of Masonite they had salvaged from one of the old military buildings. As they made their way through the deep snow toward the 300-foot crest, I heard Milt open the porch door and kick off his boots. He came up beside me and lifted the second pair of binoculars to his eyes.

"That looks like fun," he said.

"I hope it was worth the long walk," I laughed. "It must be a half a mile up there."

Through the steady snowfall, I focused on the small figures. Suddenly my grip tightened and the eyepiece jammed into my nose.

"Oh no!" I shouted. "They're standing up!" I jerked the glasses away and grabbed Milt's arm. "We have to do something!

They'll kill themselves!"

"They're having fun," Milt said calmly. "And you're missing it." He kept his glasses glued to the window. "Whoa! Look at that!"

Unable to help myself, I looked. With the glasses pressed against my face, I followed their streaking progress. Chuck stood in front, holding onto the rope, with Randall behind him, his arms clutched around his brother's waist.

They must have waxed the board because it came off that hill like a shot. About two-thirds of the way down, it sailed into the air, turned lazily like a magic carpet, and spilled the boys headfirst into the snow. The board then cartwheeled twice and buried itself into the snow.

I took one look at the dark, unmoving lumps and ran for my coat. "I have to go help them," I whispered.

With my unzipped jacket flapping and without hat or gloves, I rushed for the door. Milt called from the window. "They're fine. They just got up." He waved the glasses in my direction. "Come look."

The boys were clambering through the snow toward their makeshift sled, apparently unhurt, because as soon as they dug it out, they started back up the hill for another onslaught.

"How can you be so calm?" I asked. "Weren't you the least bit concerned?" I looked up into his steady eyes. "What would it take to make you worried?"

He put his arms around my shoulders and nodded toward the window where the boys were again sailing through the air.

"If I knew there was a possibility of seeing a car come over the top of that hill, I'd be worried," he said, his arm tightening. "Or if there were a dozen other kids up there and I knew one of them had a six-pack of beer, I'd be worried."

"Oh, dear heart," I laughed, even if my voice wasn't quite steady. "You do know how to put things into perspective."

For a few minutes we stayed by the window in a loose embrace, enjoying each other's nearness and the quiet privacy of the empty kitchen. We watched the boys brush themselves off again and start for home.

Daylight was fading, and from the corner of my eye I caught

VOLCANIC SUNSET. The air is so clean in Unalaska that sunsets usually don't have this kind of color. This one was probably the result of an Aleutian Chain volcano spewing ash. There is an active volcano on Unalaska, Makushin.

the glimpse of boat light rounding the headland. It flickered like a dim beacon through the falling snow.

"Someone is going out," I said absently. "I wonder which one this is."

It couldn't be the *Akutan* because it had already gone. The day after it brought our mail, five girls came over for their promised horseback ride. Randall was in his element, saddling everyone a horse and trotting around the horse pasture with them. But the boat had called and cut short the outing, and that night the *Akutan* left the harbor.

The light moved slowly past the window and disappeared behind the warehouse. Milt lowered the binoculars. "Cattle on the beach across the bay, going after kelp." He drummed his fingers on the sill. "I hope it rains this snow off soon, or we at least get a windstorm to blow it off the hillsides."

I raised the glasses and made out the faint dark shapes strung along the water's edge. Besides getting all their salt, the stock —sheep, cattle and even horses—got much of their winter feed off the shoreline.

Fresh ribbon and bulb kelp floated in on each new tide.

Sometimes, like now, when the grass was covered with snow, the beach between high and low tide was the only feedlot on the island.

The boys knocked on the window as they trooped past, their faces flushed and happy, great chunks of snow clinging to their caps and coats. In a moment, I heard them on the porch, stomping their feet and calling us in excited voices. We went to the door.

"That was a killer ride!" Chuck said. "Entirely killer!"

"Yeah!" Randall chimed in, fishing snowballs out of his hair. "Killer."

"Good description," I agreed. "From here it looked like a suicide run." Milt squeezed my hand and I left out the lecture on safety they had heard hundreds of times.

"Looked like fun to me," Milt told them.

"Did you see us fall on our heads?" Chuck laughed. "Boy, was that *cold*...a ton of snow right down our necks!"

"And guess what?" Randall interrupted. "We saw fox tracks!" He gave me a beseeching look. "Tomorrow is Saturday. Can we get out of school just this once so we can work on our trap line?"

"Well," I said, "maybe we can finish in half a day."

We "went to school" 6 days a week because we were interrupted so often with ranch work that we were always behind schedule.

"Oh, goody!" Randall rubbed his hands together. "I'm going to set a trap right where the tracks went under the fence."

Chuck's head whipped around. "But that's where I planned to set mine!" he protested. "You said you were going to try by the creek."

Randall stuck out his lip. "I have six traps so I can put some both places and at the Little House Creek, too, and the head of the horse pasture, if I want."

"Just a minute," Chuck growled "You can't have all the good spots." He glared at Randall and turned to us for support. "Can he, Mom?"

I'd seen this conflict coming and, short of drawing straws for the prime spots, I didn't know how to resolve it. With the competition keen between them on trivial matters, I sensed some real confrontations ahead on something as important as territory.

"Well, I…"

"We'll talk about it after supper," Milt broke in. "It's time for chores."

While I fried lamp chops and mashed potatoes, I tried to think of a fair system of sharing the trapping sites, one that could be easily enforced. I sighed, envisioning the job of policeman, knowing from past experience they could never agree on anything. I wasn't looking forward to the evening ahead.

At supper, no one talked much. Randall was our mealtime conversationalist, but he and Chuck were busy trying to glare each other to death. I played with my food and considered suggesting they take turns choosing the spots they wanted most. But that wouldn't work—they wanted the same ones.

After the dishes were washed, we all gathered around the table.

"Okay," I said, opening the discussion, "who wants to go first?"

Chuck and Randall immediately started hollering across the table at each other. I closed my eyes and started counting to 10. Before I reached 3, Milt's quiet voice cut through the squabbling.

FREE FODDER. The sea provides, in more ways than one. Kelp and seaweed washed ashore make nutritious munching for these horses, as well as the cattle the Holmeses raise. The horses the family uses were once roaming wild and had to be caught and broken.

"That's enough."

Their heads jerked up. Milt sounded pleasant and calm, and final. There was immediate silence around the table. Both boys gave him their complete attention.

"Here's what we're going to do," Milt said, looking at all of us before continuing. "First, the problem is you both want the same spots and that is impossible, so I've divided the area into two sections with Bishop Creek as the middle border. One of you will have the Little House Creek and all of West Point and the other will have Bishop Creek and Cutter Point." He paused and watched their faces.

The boys absorbed the information and came to the same conclusion. Whoever had Bishop Creek would have three of the choice sites while the boy with Little House Creek would have only one.

"I want Bishop Creek!" they said in unison.

"I know," Milt said. "And the only fair way I can come up with is for you to flip for it." He went to the desk and got a quarter out of his wallet. Flipping it into the air, he said, "You call it, Chuck."

With a groan, Randall put his head in his hands, "Oh, no."

"Tails," Chuck said.

Milt caught the coin and laid it on the back of his hand. He showed it to Chuck.

"Heads." Disappointment filled Chuck's face. "Now what?"

"Randall gets first choice," Milt said.

Randall lifted his head, a huge Cheshire grin spread over his face. "Bishop Creek," he chortled.

Although I sympathized with Chuck's disappointment, I was relieved at the outcome. Chuck was older and his salary from the crab pot business was much better than what Randall made helping with the regular ranch work of riding and butchering.

And Chuck was better at handling disappointment. He stood up and faced Randall's satisfied smirk across the table. "If you say one word," he gritted out between his clenched teeth, "just one word."

But he didn't vocalize his intention, simply turned and stalked

QUIET MOMENT. *Always calm and methodical, Milt's confidence and ability to handle any situation is the rock upon which the Holmes family is built.*

outside, his new L.L. Bean's clomping in high dudgeon.

"I don't know why Bishop Creek is such a favorite," Milt said later, after we were in bed. "I always liked Little House Creek myself. I had a fox there most every day."

"Let's hope they both have good luck," I said.

"They might lose interest after catching one," Milt predicted. "There's considerable work involved, if you do a proper job, and the whole business is smelly and messy."

"Chuck, maybe, but not Randall," I agreed. I knew how determined and goal-oriented Randall was. Nothing swayed him when he was set on a course. Even now, I bet he was in his room with his head stuck in his trapping book. If he devoted half as much time to his schoolwork, he could graduate with honors from the college of his choice.

I yawned and blew out my lamp, smiling to myself in the darkness; not over Randall's stubbornness, but the way Milt handled the problem. Not once had he raised his voice and not once had the boys even considered disagreeing with him.

Why not? Was it because he had been fair and reasonable and impartial? No, he hadn't been. In fact, he had laid down the law like a king.

Was it because they sensed he cared about them, even loved them? I didn't think so.

They knew I loved them and it didn't make a bit of difference; they still argued and disagreed with every decision I made.

Why then?

Milt drew people and animals like a magnet. Without effort, he commanded their respect and admiration. Whatever it was, he seemed entirely unaware of it.

Magic, that's what it was, pure and simple. Magic!

I snuggled my face against his magic shoulder and fell asleep.

Chapter Ten

FOOTSTEPS

By 10 o'clock, I realized I was wasting my time trying to keep the boys' attention on their lessons. As daylight increased and the first rays of sunshine fell across the table, they gazed out at the approaching day with undisguised longing.

After another half hour, I blew out the kerosene lamp we were reading by and went to the window. It was going to be a beautiful day—crisp and clear.

The snow had stopped sometime during the night and now sunlight sparkled on the unbroken white surface that sloped into the horseshoe of bright blue water of the bay. I heard a big sigh behind me as one of the boys opened a book.

ON A CLEAR DAY. The ranch headquarters sparkles a lot in winter because there's less fog when snow is on the ground. The snow usually comes in late February and melts on the ranch end of the island before it does up at the Unalaska village end.

"Okay," I gave in. "School's out."

They were gone before I could turn around. I watched them streak toward the barn, bareheaded, without mittens, and coats open to the raw chill. I knew they didn't feel it.

Milt appeared in the warehouse doorway facing the barn. The boys ran down the plank walk and joined him. After much waving of arms and pointing, they all disappeared into the barn.

I left the window. A school holiday for the boys meant one for me, too. The next few hours stretched blissfully before me. The outside sunshine beckoned. Peep-Sheep bleated from his basket.

On the other hand, the house was wonderfully warm and quiet. My spinning wheel waited in one corner. This was a perfect time to practice treadling without any interruptions.

Peep bleated again and my mind was made up. "Come on, Peep-Sheep," I said, "we're going for a walk."

I took the lid off his basket and he scrambled out and immediately began rooting at my knee. "I know you're hungry," I told him. "Just give me a minute, okay?"

I walked to the sink and Peep followed, suddenly stopping to make a big puddle on the floor. "Oh, Peep," I scolded, reaching for a rag, "couldn't you wait till we got outside?"

Undaunted and entirely without apology, Peep let loose with a clatter of little "raisins". "That does it!" I said. "As soon as you're big enough, you're going to the barn."

I put Peep back in his basket and closed the lid before he got into real mischief. After cleaning the floor, I fixed his bottle. We had switched him to instant milk; condensed was too expensive. He sucked greedily and butted his head against my hand.

Outside, Peep gamboled down the shoveled walk. But out in the barnyard, where the snow was deep, he got stuck. After extricating him a time or two, I gave up and carried him.

But even for me the going was hard. I hiked up behind the house to the cutting corrals, put Peep down in a protected corner and brushed the snow away from a patch of grass. Pulling up some, I held it out to him. "Here," I coaxed, "this is sheep food."

Peep ignored it, choosing to chew on my jeans instead. "I don't know about you, Peep," I said. "You need to learn to be a sheep."

But Peep was content to be a person and hunkered down beside my foot, with his head on the toe of my boot.

I leaned against the corral gate and gazed around me. From my vantage point on the hill, I could see the whole headquarters sprawl. On the right was the ranch house/ bunkhouse/storehouse complex. Below me on the left were the barns

COME TO MAMA. Ever since the orphan Peep-Sheep won the family's hearts, bum lambs have been welcomed by Cora in the ranch house's warm kitchen.

and the warehouses, with the small foreman's house down the beach about a quarter mile.

The tiny cottage was the original ranch house when the headquarters had been located at Cutter Point. It was moved to the beach when the present house was built and sat empty now. With the boys and me for help, Milt felt we didn't need to hire anyone.

Milt said it took a special kind of person to stay long enough to learn the vast range, with all its elusive trails and innocent-looking bogs. Usually loneliness overcame them before they became good enough to be any real help.

The foreman's house was a cozy little place, with a rushing stream and an old native steam bathhouse right outside the kitchen window. From the front window, you could look right out into the bay. When the tide was high, it almost felt like you were on a boat, the water was so near.

Birds, seals and otters played close enough to watch with-

SNOWY OVERLOOK. There's nothing like a dusting of snow to set off the creek, bay, ranch house and outbuildings on a view from a nearby hill. But no matter the weather—rain or snow, clouds or sun—the chores must be done.

out binoculars. Eagles perched on the bluff behind the steam bath, and the rich reef running along the shoreline fed a constant stream of foxes. That was why Milt thought it was such a good trapping location.

As I watched, a small figure emerged on the trail along the beach from the barn. It was Chuck carrying his traps over his shoulder, along with a motley collection of beef entrails he had collected from the slaughterhouse for bait.

I glanced in the opposite direction and saw an even smaller figure trudging through the snow toward Bishop Creek. Even at this distance, I could see Randall was bent nearly double under his load of equipment and bait.

A shovel stuck out from under one arm and a length of chain dragged in the snow. It looked like Randall was going to try every set he had read and told me about; from the beach type to the tundra one.

The day was too nice and I was too restless to go back inside. Tucking Peep under my arm, I hiked through the fenced pastures until I reached Sockeye Lake, a mile from the headquarters. It lay below me like a sapphire nestled in a deep box of snow-white cotton, quiet and undisturbed.

Around the lake's edges, a ring of black sand stretched, bare and smooth, where the lapping waves melted the snow. I put Peep down and jogged back and forth on the sandy beach until I felt perspiration trickle down my back.

As I ran, I tried to compose a letter to my sister Alice; something I had been avoiding because I didn't know what to say to her. Her husband was dead. What could I possibly say that would comfort her? Nothing.

When the creek that fed the lake on its eastern end stopped my progress, I turned around. As I jogged back, I saw my tracks leading away into the snow—a set of single boot prints, meandering and solitary.

The tracks made me think about a prayer I had read somewhere about a man who had dreamed about talking to God. The man had questioned why the footprints were single, when God promised never to desert him. God replied that the single set of footprints were made when God carried him.

Was Alice feeling deserted? I didn't know how strong her faith was, but I was going to find that prayer and send it to her, along with a letter telling her how sorry I was that Basil was gone.

Trite? Maybe, but 100 times better than saying nothing at all. She would at least know that I cared, even if every word I wrote was wrong. I scooped up Peep and set off toward home with a lighter heart.

As soon as I reached the house, I dug through the magazines until I found the prayer, then I sat down and wrote the letter. When the letter was tucked into the mailbag, I poured a fresh cup of coffee from the pot simmering on the back of the stove and pulled my spinning wheel closer to the hearth.

I had been reading a spinning instruction book even more carefully than Randall did his trapping manuals, and I'd decided that when it said to practice treadling without spinning yarn, it meant me. Always eager to get to the actual deed, I had skipped all the boring stuff about how to make the wheel work. Now I sat down and put my foot on the treadle.

With a lurch, the wheel snapped backward, stalled and came to a standstill. Instead of jumping up and down on it until I had reduced it to kindling, which was my first impulse, I took a deep breath and looked again at the book.

Rereading the text and studying the pictures, I put my hand on the wheel and set it gently in motion before I attempted to treadle. When I had it whirring smoothly, I eased my foot onto the treadle.

Whap! The wheel lurched backward and jerked to a stop. The

YARN SPINNER. The spinning wheel managed to elude all of Cora's efforts, until she made a simple adjustment. Now piles of wool are turned into skeins of yarn and Cora skillfully knits warm garments and other items for the family.

adage about some people not being able to chew gum and walk at the same time floated through my mind. A suspicious wetness built behind my eyeballs and I sniffed.

This was ridiculous. I refused to weep over such a silly problem, and I refused to be defeated by a function that every 10-year-old girl in America could perform without pause 150 years ago. It was a matter of pride.

Okay, I couldn't make the wheel spin. Something was obviously the matter. What was it? I stared nervously at the spool and pulley arrangement perched in front of the wheel. It was called a flyer and was supposed to do the actual making of the yarn from wool.

Somehow I had the idea that if a handful of wool was held somewhere near this complicated series of wheels and bobbins, it would snatch it in some mysterious way and turn it into yarn. So far, that hadn't happened.

Far down on the right side of the flyer was a beautiful little knob with a line attached to it. This line ran through an eye and up over the spool that held the finished yarn and down the other side to where it tied to another eye. The book identified this as the tension device.

Maybe that was it. Maybe the tension was wrong. I loosened the knob. After all, what could it hurt? And I had tried everything else.

Again I set the big wheel in motion with my hand, then ever-so-slowly pressed my foot down on the treadle. It was like stepping into soft butter—no resistance at all...no lurching...no jerking...and no spinning backward on the big wheel. I hardly dared breathe. Could the solution be this simple?

Up and down, up and down. Bravely, I lifted my hand from the wheel. It continued its smooth forward motion. Oh, this was bliss! I pedaled softly—up and down, up and down—spinning. I was actually spinning! Well, not really. But I was making the wheel work.

I closed my eyes and let the soft whirring of the wheel wash over me. Mindless and pleasurable. I had read that somewhere. Spinning is a mindless and pleasurable task. For the first time since I had started my struggle with the wheel, I caught a glimmer of what that meant.

For a minute, I let myself dream. The kitchen was warm and quiet, with winter sunshine filtering through the windows to lie in bright patches on the worn linoleum. In my mind's eye, I saw myself in long skirt and dust cap, feeding wool into the flyer and watching it emerge as yarn on the other side, to pile

HEAT SEAT. Randall and Cora find a warm seat on the cookstove door in the kitchen. It's a cozy spot that over the years has served as seat, footrest and even an incubator for an orphaned lamb.

in great heaps around the legs of my stool...

The outside door slammed. "Mom, help me!" My eyes popped open and I knocked over my stool getting to the door.

"What's wrong?" I recognized Randall's voice but he sounded strange. I pulled open the kitchen door. "My God, Randall, what happened?"

He stood in front of me, a sodden, shivering bundle of dripping clothes and pale, clammy skin. Even his cap and hair were wet.

"I-I f-fe-fell in the creek," he stuttered. His lips trembled. "And I l-lo-lost two tr-traps."

"You can tell me later," I interrupted. "Get out of those wet things; clean to your skin." Dashing into the bathroom, I grabbed a towel. "Here, wrap up in this." I stuck it through the door. "And come by the fire."

Randall scampered in and stood shivering in front of the ash box, water still dripping from his hair and splashing on the hot stove surface. I opened the oven door and a blast of hot air enveloped him. Taking one end of the towel, I rubbed his head vigorously until he jerked away.

"Stop wiggling," I scolded.

"You're pulling my ears off," he complained.

"I'm just getting you dry." That wasn't exactly the truth. I was holding him and touching him; reassuring myself that this sturdy little body was alive and whole. I gave him a final swipe.

"Now, how did you fall in the creek?"

"Can I get some clothes on first?" He held the wet towel away from his chest.

"Okay, but hurry. I want to know what happened."

I moved the spinning wheel back to the corner, unable to concentrate on treadling while I was so shaken by Randall's accident. Getting wet was an occasion for concern; something even Milt was stern about. It could have serious consequences under the right conditions; like a good, brisk wind and a distance to travel. Hypothermia was always a near and present danger.

I fixed Randall a sandwich and heated some tomato soup. I hoped he had learned a lesson, whatever had happened. However, when he came into the kitchen, he looked neither penitent nor subdued.

"Man, that water was cold." he sat down on the cool oven door and took the bowl I handed him. "Brrr!"

"I'm listening," I said.

He shrugged his shoulders and bit into the sandwich. "I was walking along the creek, going to the head of the horse pasture. I thought I was far enough back from the edge, but the snow went right out over the water and I walked up on it and fell through."

He gave a little shiver. "It was deep and cold. I dropped the traps and they sank."

I knew the creek was high from all the snow. Wanting to somehow impress upon him the gravity of what had happened, I said, "You could have drowned or died from hypothermia."

"Mom." He looked at me with exaggerated patience. "I fell in the creek and I jumped right back out. Then I ran home as fast as I could. I didn't have time to catch hypothermia."

"What if you had been farther away," I persisted.

"But I wasn't," he said with a puzzled frown. "I was real close." He stuffed the last of the sandwich into his mouth. "You worry too much." Handing me the plate, he stood up.

"I wonder if Milt has any more old traps." He escaped before I could scold him further.

But I wasn't finished. That evening, after supper, when we were all in our usual places—Milt and Randall pulled up to the stove with their feet in the oven, and Chuck in a chair beside the stove with his back against the warm hot water tank—I asked Milt about the danger of snow camouflaging the creeks and what the boys should do about it.

Milt looked up from his livestock journal. "Randall told me about his adventure," he said. "I told him to carry a stick with him next time so he can test the snow before he walks on it."

Randall looked at me with a complacent smile. So simple and so effective, and something I wouldn't have thought of if my life depended on it.

"Good idea," I mumbled.

The kitchen was quiet. With the generator providing electricity to read by, the boys and Milt stayed absorbed in their magazines until time for late chores.

I took advantage of the power to do laundry. Without a dryer, I hung the clothes on lines strung across the utility room adjacent to the kitchen and warmed by the big coal stove. In 24 hours, they would be dry and I could hang up another load.

Peep-Sheep loved it. When I let him out of his basket, he took two steps from the washing machine to the clothesline with me and then back to the machine, keeping perfect time, always right by my knee. I was his mother and he wasn't losing me.

I had just washed all the wet clothes Randall had left on the porch. As I shook them out and pinned them to the line, I heard him say to Chuck:

"It is so true—you just ask Milt."

Chuck straightened in his chair. "Milt, did you tell Randall we could get $50 for a fox hide?"

"If you do a good job of skinning, fleshing and stretching them, yes," Milt answered. "I took mine to the Fur Rendezvous in

COLD CUTS. *The snow is never so bad for so long that the horses can't find some frozen fodder. The range animals at Chernofski have to be of hardy stock, as the only feed they get is what they find in the various pastures and along the beach.*

Anchorage and got at least that much at the auction." He folded his journal and laid it across his knee. "But it's a lot of hard work. I did nothing else all one winter while I was here alone and only got 60 pelts."

"Sixty pelts," Randall breathed. I could almost hear the cash registers behind his eyelids, tallying up the figures.

There followed a discussion on pelt preparation that I only half listened to. While Milt explained the finer points of skinning out ears and feet, I fed Peep and sort of hoped that fox trapping was one of those endeavors that would peter out after a few unsuccessful attempts. It sounded complicated and messy and involved sharp tools.

Skinning knives, fleshing blades? They were talking some serious stuff.

"It's 10 o'clock," I broke in. "Time for bed."

The next morning at breakfast, Randall was very quiet. Neither he nor Chuck had caught a fox during the night and I was quietly encouraged. Randall's next question lightened my heart still further.

"Milt, did you ever catch a colt from the west side?"

Milt looked up from his pancakes. "My horse, Fuzzy, was a wild colt."

"How did you catch him?" This from Chuck.

"There used to be a big old stud with a bad hind leg," Milt said. "He couldn't run fast, but the mares wouldn't leave him. A bunch of us went down there and drove his herd to the ranch. We cut out Fuzzy and Poncho and a couple others."

"Could we do that now?" Randall asked. "I want my own colt."

"Well," Milt scratched his jaw, "that old crippled stud's been dead a long time. I don't think we could get close to anything else with the horses we have now. Fuzzy is 16 and he's the youngest."

Randall was not to be put off, however. "If you wanted a colt right now, what would you do?"

"Hmmm. I don't really know." Milt stirred his tea. "I guess I'd wait until about March when the passes are all drifted in and the horses thin from winter, and I'd try to run a bunch into a drifted pass or a bog, then rope a colt and lead him home."

"Let's do it," Randall urged.

Milt shook his head. "Even with the wild horses thin and the passes drifted, we don't have anything that could keep up with them. Our horses would get bogged down, too." Milt pushed back his chair. "Besides, we have enough horses."

Randall rested his chin on his hand and stared out the window. "What if I had a motorcycle? I bet I could catch one then."

"Yeah, maybe you could at that," Milt grinned to himself and shrugged on his coat.

Long after everyone had left the table and I had the dishes cleared away, Randall sat scribbling furiously on a sheet of paper. Finally I leaned over his shoulder to see what he was doing. A column of figures straggled down the page.

"What are you doing?" I asked. It looked like he was adding a whole string of the number 50. "Why don't you just multiply?"

"Because I can never remember how," he said impatiently. I watched his fingers drum on the table and knew he was counting on them.

"Let me show you," I offered.

"No," he said. "Leave me alone, I've almost got it figured out."

Numbers were hard for Randall—almost impossible for him to grasp. The fact that they were hard for me, too, didn't make it any easier for him.

"Why are you adding all those 50s?" I asked.

With a flourish, he scribbled a figure. "I want to know how many foxes I need to catch for enough money to buy that motorcycle Chuck and I saw in the catalog."

Triumphant, he showed me the tortured page. "I only need 12." He wadded the paper and made a perfect basket in the coal bucket. "Then I'll catch a wild colt."

After he left, I swept the floor. Motorcycle? Never! He could really hurt himself on that. Thinking about a wild colt was bad enough.

Still, I wasn't really worried. He hadn't even caught one fox yet. And how could he get a motorcycle up here when it was a fight just to get mail. I hummed as I stoked the fire and started to make cookies.

I should have known better.

WOOLLY PARADE. Chuck and Randall, in background in top photo, bring in the sheep for the annual tick-control dip. The sheep are sheared about a month before, which allows their fleece to grow out enough to retain the solution. Above, orphaned sheep get a real treat—oats. Only the orphans get spoiled like this, and it makes their fleece strong and beautiful. The range animals get no supplement in their diets and don't know what oats are.

Chapter Eleven

A HARD ONE TO LOSE

Before I got up and lit the fire the next morning, I heard Randall and Chuck leave the house. I listened to the wind whistle outside our bedroom windows. I hoped they had their coats zipped, but I doubted it. They had trapping in their blood; nothing else mattered.

While I was fixing breakfast by lamplight, they returned empty-handed.

"Two traps set off," Randall said glumly. "And my bait gone." He dropped his fox-whacker on the floor with a thud.

"Well, you know the old saying, 'Wily as a fox'," I sympathized. "They won't be easy to outsmart." I turned to Chuck. "How about you. Bait gone, too?"

"Nah." He reached behind me and stole a slice of bacon. "Not even any tracks, except birds." I slapped his arm automatically but he got the bacon into his mouth and grinned at me around it. "I'm not going to catch a fox there—it's a dead hole."

"Give it a chance." Out of the corner of my eye I saw the end of a bacon slice disappear behind Randall's teeth. "Come on, you guys. Cut it out." I grabbed the plate and set it on the warming oven.

"We're hungry," Randall said.

"Then go wake up Milt while there's still some breakfast left." I broke eggs into the sourdough and added sugar.

While we were eating, the processor *Galaxy* called on the VHF. They passed on a message from Standard Oil in Dutch Harbor. Charlie Brown in Homer, Alaska wanted to know when the mutton he had ordered would be coming.

"As soon as I can line up some transportation," Milt answered. "I don't want to butcher until I know they have a ride to town where they can get on a plane."

"We're leaving in about 5 days," the *Galaxy* skipper said.

131

FLAPJACK FLIPPER. It may be an old coal-burner, but the kitchen cookstove turns out great meals, as Cora proves with pancakes for a hungry crew. The kitchen is the family's gathering place, not only for meals, since the stove provides welcome warmth.

"I'll be glad to take it in for Charlie if you can get it over here."

"That won't be a problem," Milt assured him. "We'll get them over there in plenty of time—and thanks."

Immediately, everyone's plans changed. Even though a northwester was blowing between 40 and 50 knots, we all hiked a half mile to Pasture 5 and brought the sheep down to the barn, where we separated out 15 fat wethers.

As we herded them into a small side pen, Milt said, "We'll leave them to gaunt up for a day; let their systems clean out. We have plenty of time." He slid the panel closed behind them. "We'll start butchering day after tomorrow and do five a day."

On the way back up to the pasture with the remaining flock, we saw a monster surf breaking across the shallow reef in front of ranch headquarters. Riding on the waves like insane surfers were three huge driftwood logs.

As the waves broke, the logs canted skyward through the foaming spray like matchsticks. They were smooth and round, without limbs. I paled. Saw logs, probably washed off a barge going to Japan. Please, God, don't let there be any more.

During a northwester, the marine railway tracks were unprotected and vulnerable to floating debris. The boat dolly with our dory was lashed to the tracks and winched up to its highest point, over dry land and out of danger. But there was no way to get the tracks and supports out of the water or protect them.

Milt was already running. He yelled at Chuck and they both plowed downhill through the heavy snow and vanished into the barn. As fast as we could, Randall and I followed the sheep to Pasture 5.

By the time we got the gates closed, Milt and Chuck were scrambling across the gravel beach, armed with ropes and pike poles. Neither of them had on rain gear, and blowing spray drenched them in an instant.

Randall and I ran, too, and as we got closer, I saw many small logs, limbs and mats of ropy seaweed pounding on the shore. The wind was so strong it was impossible to look directly into it. We sheltered our eyes with upflung arms and ran on.

When we reached the barn, Randall grabbed a pike pole and bolted for the door. "Go to the house and get on rain gear," I ordered.

"But Chuck doesn't ha..."

"Don't argue," I snapped. "There isn't time."

I grabbed his shoulder and pushed him toward the house. Inside, he jerked on a pair of rubber bib overalls over his clothes and I made him stand still as I fastened the hood of his slicker around his face. Then he was gone.

Before I struggled into my own rain gear and high gum boots, I filled the firebox with coal and closed the damper. It might be hours before any of us came back.

Forty feet from the beach, the salt spray hit me, stinging my face with wind-driven needles. Long ribbons of seaweed whipped into the air. As I got closer, the noise of the surf drowned out every other sound.

I saw Randall staggering and crawling over rocks toward Chuck and Milt. They huddled midway down the beach with a pile of coiled ropes at their feet and pike poles clutched in their hands. With legs spread wide and their bodies bent nearly double, they still lurched sideways and grabbed each other

PILES OF POTS. The storage of crab pots was one source of income for the Holmes family. The Bering Sea is a crab-catching paradise, and the ropes and rigging piled near the pots are handy in this maritime occupation.

for support as the wind hit them.

I crouched on the bank and stared at the wall of water with its cargo of destruction as it screamed toward the shore. The logs were broadside, carried along like toothpicks, and dashed parallel against the beach with each surge, then pulled back out with the churning undertow.

It looked hopeless. The logs were immense and each wave hurled them nearer the marine railway. Ducking my head against the spray, I scrambled down the bank onto the beach in time for Milt to push the end of a rope into my hands. Randall hurtled into my side and added his hand to mine just as Milt shouted, "Now!"

As the first log hammered the beach, Chuck drove his pike pole into it, 10 feet from the end. He jammed the top of the pole into his midriff and clutched it with both hands. With his cheeks beet-red, he pushed the log back into the water a few inches.

From where I stood, I saw a look of pained discomfort streak across his face and I knew that his boots had filled with icy sea water.

"Got it!" he gasped.

Milt ran in close to the water and lassoed the free end of the log. Rapidly working the rope down until the loop rested against Chuck's pike pole, he jerked it tight.

"Pull!" he yelled.

Randall and I put all our weight against the rope and felt the log bump shore again, Chuck jerked his pike pole free and helped us. When the wave went out, we dug in our heels and fought against the sea for the inches we had gained.

I twisted so the rope rested around my waist and it cut into my side. I glanced back at the other three, oblivious to the spray cascading over me. Their faces showed the same strain and discomfort I knew mine did.

None of us let go.

Hurry, hurry, hurry.

The next log loomed closer. Already it had passed us and was making progress toward the railway.

Another wave and we gained a foot. As we battled the undertow, Milt came hand over hand up the rope and took the end from me.

"Far enough."

He leaned close to Randall and shouted, "Get the Ferret!" Then he snubbed the rope around a gnarled driftwood stump just as the wave went out. The rope stretched like a bowstring, but the stump didn't move.

Chuck struggled down the beach, staggering sideways, doing a slow crab crawl when I knew he was trying to race. He watched the wave through the stinging spray, then impaled the next log for Milt to rope.

We strained to bring it up the beach and tie it off. We were just finishing the third one when we heard the Ferret engine over the roar of the surf.

Good. I wiped water out of my eyes. Soon Milt and Chuck could go to the house and warm up. They must be freezing. My toes were numb and I had stopped feeling my hands after the first log. But I was dry under my rain gear.

The Ferret made short work of dragging the logs past the high tide line where they wouldn't be a threat to the railway.

I clambered over the slippery beach gravel and stepped up

on the bank. Milt and the boys were coiling the ropes, their backs to the wind and the sea. Tugging on Milt's arm, I shouted, "Go get into dry clothes. I'll do this."

Milt grinned through the water running down his face. "Easy firewood, huh?" I made a face at him and motioned toward the house.

"Chuck!" Milt hollered. "Mama says go change your clothes." Chuck didn't need any urging. He finished his coil and started for

ALL-WEATHER GEAR. It could be cold. It could be wet. It could be both. Whatever the weather on Unalaska Island, the Holmes family has the gear for it. Coats, jackets, slickers and rainsuits are all hanging handily by the door.

the house. With the wind at his back, he had to hold himself rigid to keep from being blown over on his face into the snow.

"You, too," I insisted, glaring back at Milt from under my hood. "You're soaked."

"Milt!" I heard Randall shriek. Both of us whirled around. A log was in the railway!

A gnarled, blackened snag, water-logged and heavy, carried just under the surface of the water, had swept down the beach, unnoticed while we struggled with the floating logs. The first third of it was riding free between two sets of pilings, with its long end whipping like a snake and crashing with battering ram force into the beleaguered uprights.

"Hurry!" Milt stumbled over the rocks. "We'll have to push it on through," he yelled.

Like stiff robots, Randall and I picked up the pike poles and followed. Maybe it wouldn't be too bad. I held the pole low so the wind couldn't wrench it away from me. All we had to do was get the log lined up with the opening and push on it. The water would do the rest.

But when I got to the tracks and crawled up on the rail, I saw that pushing would be impossible. Milt was already braced on the crosspiece above the jammed log, his face a set mask. Randall was perched on the far track. Both of them had their poles wedged beside the log.

When the wave receded, I glimpsed a broken limb extending from the log under the crosspiece Milt stood on. It was caught against the far piling and no amount of pushing with pike poles would dislodge it.

Each new wave slammed the log into the upright, where both the log and its limb exerted force. Even the strongest piling couldn't withstand that kind of force for long, and the railway had been built nearly 20 years ago.

I slashed my hand across my throat, signaling Milt to stop. "A branch!" I shouted, pantomiming a tree crotch. "It's caught."

I made sawing motions. Milt inched his way to where I stood. His eyes took in the grim spectacle. He repeated the sawing motion and pointed to the winch house.

His lips were blue, he shook with cold. I wanted to make him go warm up. But I knew insisting was useless. He wouldn't leave until the railway was out of danger. I went for the saw.

Coming out of the winch house, I bumped into Chuck, his crouched body bright in orange rain gear. Our heads close together, I explained the log and handed him the saw.

"Don't give it to Milt," I begged. "Make him show you what to do. He's been wet too long."

Chuck nodded. Together we forced our way back down the bank.

"Milt," Chuck yelled, his face creased in a big grin, "Mama

says go change your clothes." Milt reached for the saw. Chuck pulled it out of his hands. "Show me what to do."

Milt hesitated. He took off his gloves and wrung the water out of them, sticking his hands inside his coat, under his arms.

"Okay," he said. "Get in here where I am. Wait till the water goes out and start sawing from this angle." He glanced up at me. "Mom, you and Randall try to steady it with your poles."

He inched his way off the crosswalk. "If you get it sawed off before I get back, wait to push it out until I come." He sidled through the water on the crosspiece, slung his legs stiffly over the rail and stumbled up the beach.

Chuck lowered himself into the knee-deep water and jockeyed the saw into position above the thick limb. Above him, Randall and I pushed against the log with our poles.

The limb was submerged at least a foot. It was hard work to maneuver the stiff saw under water. I watched it bow and buckle as Chuck tried to get a purchase on the sticky wood.

The next wave overtook him after only two successful passes. It swirled around him waist-high and its spray covered us all. We waited for it to recede, grimly holding our positions. When it was gone, Chuck sawed furiously. I had counted 16 waves when Milt scrambled down the beach in his own fluorescent rain garb.

The limb was just half-severed. Milt took the saw from Chuck and pumped his arm with piston swiftness. Then he stomped his foot into the water with a huge splash and the limb snapped.

"Push!" Milt's voice cracked—hoarse from shouting directions above the noise of the water. "Push!" I leaned into my pike pole. With every ounce in my body, I willed that log to move.

I felt its sluggish motion. Like a huge lizard coming awake on a sun-warmed rock, the log began to move; slowly at first, then with increasing momentum. I sensed Chuck behind us, adding his strength to ours.

Suddenly it was done. The ragged end slid past the piling and floated free, rolling and bumping in the tangled seaweed as the waves pushed it farther around the beach, away from the railway.

We watched it stupidly, as if we couldn't really believe it

ALL LOGGED IN. The huge saw log that almost wrecked the marine railway comes to a useful end as Milt saws and splits it into firewood. With no trees on Unalaska Island, logs washed ashore are always welcome; if they don't hit the boat launch!

was gone. Milt crawled out of the water to stand beside us. I could hear the water sloshing in his boots. "Good job," he said.

"Milt, Chuck and Randall!" I shouted one last time. "Mom says go change your clothes."

We took one final look at the foamy, rushing water, decided none of the debris in it was a threat, and made a dash for the house. The wind pushing us didn't seem as strong as when we started out. Maybe we were just used to it.

Back inside, we changed our wet clothes. We drank cups of hot tea, ate cold pancakes rolled up with brown sugar and basked in the kitchen's warmth. Every few minutes, one of us would go to the front window and scan the water with binoculars, watching for another log that would send us headlong for the beach.

"I don't think it's blowing as hard as it was," Chuck remarked while he peered through the glasses. "But those waves are still humongous."

I leaned my elbows on the sill and stared at the wind sock

whipping above the gatepost. "I think you're right," I agreed. "Peggy said it was supposed to swing around to the southwest tonight."

"Let's hope she's right," Milt spoke from the stove. "We need a good southeast rain to melt this snow off." He set his cup in the sink and picked up Peep-Sheep, who was nibbling on his jeans. "You want to come with me?" he asked the wriggling lamb. Tucking him under his arm, he said, "I'll be in the barn working on saddles. If you need me, ring the bunkhouse bell."

When the outside door closed behind him, I said, "Back to normal, boys. Get out your books."

"Aw, Mom, do we have to?" Chuck groaned. "I want to

WAVY WEATHER. *Storms like this control the lives of the Holmes family. There'll be no work on the water this day, as all hands will keep a watch for more drifting logs that can wreak havoc on the docks and boat launch.*

move my traps and go watch the waves break on West Point before the wind changes." He looked at the clock. "And it's almost 1 o'clock now."

"Yeah," Randall chimed in. "And we worked hard all morning. We deserve a break."

"Don't even consider it," I stated in my firmest tone. "We have this afternoon and tomorrow to get a couple of lessons ahead. Then we'll be butchering sheep and the week will be gone." Reluctantly, they complied.

In the 3 hours that they studied, the wind dropped steadily. I watched the bay through the glasses almost continuously, afraid if I relaxed my vigil, something huge would sneak past me and lodge itself between the pilings again. We might not be so lucky the next time.

At 4, I dismissed the boys and they streaked away. The wind sock fluttered gently and rested against the pole. Except for the huge breakers still crashing into the reef, no one would believe the violence of a few short hours ago.

Belatedly planning my supper menu, I took a saw to the warehouse. Outside the air felt much warmer, almost balmy. I checked the thermometer mounted on the storehouse wall—42°; up from 30° this morning.

Oh good, a chinook. I lifted my face and let the warm air caress it as I hurried over the soggy snow. I had heard many bitter complaints about Aleutian weather, and for the fishermen, they were all valid. The winds were horrible, deadly, unpredictable and killed without compunction.

But for us, the maritime climate was often kind. In fact, our Aleutian weather was about the world's best-kept secret. The temperature rarely dipped below zero. The snowfall was moderate and, following a rough cycle, usually got rained off to bare ground every 10 days or so. It made grazing possible year-round.

When the weather warmed up as it was doing now, the grass even grew. And in the hilly tussock center, there was always some green showing, tender and unprotected. There was also the beach, with its rich abundance of seaweed and salt.

In the warehouse, I sawed off a hind leg of mutton, glad that the small amount left felt icy, partially frozen. We'd have it eaten before the warm weather caused it to mold.

Carrying the leg above my head so the dogs wouldn't take it away from me, I got it to the house and into the oven. Then I picked out the least shriveled potatoes and knocked the sprouts off the rest of them.

I needed to order groceries. With school and ranch work always crowding in, I kept putting it off. I promised myself I'd do it when the boys studied.

When the boys came home at dusk, their faces were ruddy and excited. All through supper they chattered about new sets and locations. Chuck seemed to have accepted his less-than-satisfactory territory, waving his hands as much as Randall as they interrupted each other to tell us just how they brushed away their tracks, hid their bait under driftwood so the birds couldn't steal it, and sprinkled dirt over the trap pan to conceal the set.

Later, they both went to bed without grumbling, eager for the morning and the promise of success.

In the darkness I again heard them leave before I lit the fire. Again they came back empty-handed and wet through from the gentle southeast rain that had started during the night.

Right after breakfast, while I fed Peep and the boys cleared the table, the *Galaxy* called.

"Sorry, Milt," the skipper apologized. "But our orders have changed and we're leaving at 5 in the morning. Can you get that mutton to us by this afternoon?"

"Uh…we can try." Milt paused and rubbed his jaw. "We can get some, anyway."

"I hate to do this to you," the skipper said. "But I don't have any choice. The owners want the boat out of here."

"I understand," Milt reassured him. "You're doing us a favor any way you look at it." He signed off and came back into the kitchen.

"What do you think?" he asked me.

"If we do one every 30 minutes, we can get 10 done in the daylight," I said. "The *Galaxy* can put them in a refrigerated room and they'll cool out overnight."

"That many will mean two skiffloads," Milt reminded me. "And we'll be taking them across the bay in the dark."

"Then we'll just hope the weather holds," I answered. "And if we can't get them on the boat, we can always salt them down for our Norwegian customers. We've done it before."

"Okay, we'll try it," Milt decided. He jerked open a kitchen drawer and began selecting knives.

The breakfast dishes were forgotten. Classes were not even mentioned. I collected muslin and a kettle for hearts and livers. Chuck filled a stainless steel bucket with hot water, while Randall stacked a coal bucket on top of Peep's basket so he wouldn't escape and wreak havoc while we were gone.

In the barn we quickly prepared our work area. Since sheep were smaller than steers and cooled out faster, we didn't use the slaughterhouse, butchering instead in the back of the barn where Milt had a hand hoist.

For the first two animals, we used a lantern. We all knew our roles and they went smoothly. At first, skinning a sheep had been difficult because it was so different from doing a steer. Instead of using a knife, Milt pushed the hide off with his clenched hand. "Fisting", he called it.

It left the carcass smooth and glistening under a thin membrane Milt called "striffen". There were no nicks or slashes to mar the flesh and let in dirt. It also meant there was less likelihood of cutting the pelt, which, when tanned, brought more than the animal itself. Chuck soon caught on and now I couldn't tell his side of the hide from Milt's.

When the first one was finished, I washed it down inside and out with a rough cloth and steaming water. While I did this, Randall emptied the wheelbarrow of entrails on the beach,

MUTTON TO GO. Freshly cut and wrapped in muslin, this meat is being hauled by Ferret to be transferred to the Holmeses' best customers—commercial fishermen.

where the eagles and gulls devoured them before he came with the next load.

As soon as I finished, Milt covered the carcass with muslin and Chuck shouldered and carried it to the meat hooks in the warehouse. While they were gone, I sluiced the floor with the rest of my hot water and Randall hung the pelt fleece-side down over a panel.

"Are we saving the heads for Helga?" he asked. "Or can I have them for bait?"

"Better save them," I said. "We promised."

I covered the heart and liver with a clean cloth before running to the house for more hot water. The whole job had taken 27 minutes.

I added coal and checked on Peep. He was darn near as much responsibility as a baby.

When I returned to the barn, Milt and Chuck were busy on the next animal.

We didn't stop for lunch. The soft rain continued to fall as the neat, white bagged shapes added up in the warehouse. We all stayed busy, except Randall.

His tasks didn't take him long, and between he grew restive. We were working on the eighth one when he disappeared. I don't know how long he had been gone, but when it was time to wheel out the offal, he was nowhere to be found.

I stepped out of the small side door and looked around. Where could he have gotten to? I was just getting ready to holler when Milt and Chuck came out of the warehouse after hanging up the last carcass. They were laughing about something and didn't notice me.

All of a sudden, Randall galloped through the gate from the beach house pasture. Bumping the ground behind him was the biggest, reddest Aleutian fox I had ever seen.

I must have made some sound because Milt and Chuck stopped laughing and looked around. Randall skidded to a stop in front of Chuck and dropped the fox on the snowbank.

"Here's your fox, Chuck," he said with a smug, satisfied look. "I whacked it on the nose and it keeled over."

Chuck tried to speak, but a whisper was all he managed.

"What?" He looked at Randall in disbelief. "You did what?"

"Yeah," Randall gloated. "I didn't have one in my traps but you did and I wanted to try out my whacker, so I got yours for you." He stood back and waited in triumph. "It worked like a dream...just like the book said."

Chuck exploded off the walk. "Why you little..."

"Hey!" Milt yelled at the arguing boys. "He's getting away!"

Our eyes riveted on the snowbank where the groggy fox had come to, got up and staggered around.

"Better step on him," Milt warned.

Chuck leaped into the snow, one foot landing on the animal's chest. But the snow was soft from all the rain and his foot pushed the fox deep into the bank.

The icy bath revived the fox still further, and he whipped his head around and sank his teeth into Chuck's new L.L. Bean boot.

"AAAAAHHHHH!" Chuck jerked up his foot.

The fox was gone like a flaming arrow against the snow, streaking across the walk, flashing through the barnyard, sailing over the bank and disappearing from sight.

"Did he bite you?" I shrieked.

Chuck shook his head, apparently too astonished to talk.

We all stared at the quiet, empty barnyard, then we looked at each other.

"Well, that's that," Milt broke the silence with his matter-of-fact voice. "Let's get back to work." He strode into the barn.

"Good idea," Randall said, making a wide detour around Chuck and dashing after Milt. "I'll help you."

Chuck was a statue in the snowbank, his cheeks bright red, his eyes glazed.

"I'm sorry, Chuck," I attempted to console him. "There'll be more foxes. Don't worry."

But Chuck wasn't listening. "Just like my fishing pole," he whispered through clenched teeth. "That kid gets away with murder." His eyes blinked furiously.

"What would you like me to do?" I asked reasonably. "Randall didn't mean to lose your fox. He thought he was doing you a favor." My lips twitched.

"Mom," Chuck's face worked, "if you don't do something, I will." But his voice cracked and, in spite of himself, he grinned.

"That rotten brat, I can't believe he did it." His forehead puckered in an irritated frown and he glared at me in annoyance. "Whenever I think he's done everything dumb there possibly is, he comes up with something new."

I nodded but my lips twitched uncontrollably and I snickered. "Did you see his face when that fox came back to life?"

Chuck nodded. A chuckle escaped from him.

In a moment, we were both laughing until tears ran down our faces and we gasped for breath. We couldn't stop. We had to hold each other up. I knew we should be inside helping, but all I could do was giggle.

All of sudden Chuck stopped laughing. A look of absolute horror seeped into his eyes, as if he had just realized something dreadful.

"What's the matter?" I asked.

"My foot is wet," he whispered. In slow motion, he lifted his leg out of the slush and peered at his new boot. There on the side were two neat puncture holes, right where the sole and the upper came together.

"Oh, no," I groaned. "Did he get clear through?" I knew how much Chuck loved those boots. "Maybe Milt can patch them."

"It won't be the same," Chuck despaired. "They'll be old, patched boots, and I've only had them a week." His eyes glinted with a fierce light. His lips tightened into a determined line.

"Now, Chuck," I cautioned. "Don't do anything rash."

"I'm gonna pound him," Chuck snarled. "Right into a pulp."

During the next 2 hours, while we finished the sheep and got them ferried across to the *Galaxy*, Chuck maintained a ferocious silence and Randall stayed very close to Milt.

I don't know what happened between them later. No bruises were ever apparent. But I know it's safe to say that any fur flying certainly didn't belong to that fox.

Chapter Twelve

A LONG WAY FROM MY HEART

Gosh, Milt," I called as he came through the gate from the chicken yard, "can you believe this weather? It's January 20 and we're having another beautiful day."

I came out of the storeroom with my apron full of vegetables from the sand barrel. "Sunshine, blue sky and 30°. No one would believe this is the frozen north."

Last week we hadn't been so lucky when we tried to load beef on the *Tamar*. The day started as calm and beautiful as this one. Before we had one wheelbarrow loaded, however, a sudden southeast storm of about 80 knots blew up and wedged the 100-foot boat between a right-angle jut in the dock and the rocky shore.

Getting it free had been a tense struggle for everyone. While Milt and Chuck were rigging a springline from the *Tamar's* bow to a timber far out on the rickety dock, trying to give the boat something to push against, another boat appeared out of the storm and was blown into the dock. Timbers and planks shot skyward and both Milt and Chuck disappeared into a sea of spray and foam.

From the beach it looked like they'd been washed away. But when the water receded, I saw glimpses of fluorescent rain gear and knew they had survived.

Crawling around the holes, they had picked up the springline and secured it so the *Tamar* could forward her engines without going aground and generate enough thrust to swing her stern past the angle in the dock.

The other boat was the *Liberty*. On its way west to the fishing grounds, the *Liberty* had picked up our mail in town and was trying to deliver it. Even seasoned skippers, like those on both the *Tamar* and the *Liberty*, were helpless against such winds.

Although the storm was short-lived, blowing itself out in 12 hours, it was another 4 days before the boat returned for the beef. The second attempt went better. We loaded 20 quarters of beef and another 17 sheep aboard without a hitch.

That delivery marked the end of our winter butchering season. In the last 2 months, we had butchered 12 steers and 35 sheep. Interspersed with crab pots and class work, the long cold days of butchering and riding had taken their toll. All of us were glad to see the final shipment leave the harbor.

"It's too nice to stay inside, Milt," I said. "Let's do something."

"Sure," Milt agreed, opening the porch door for me and putting his foot across the jamb just in time to keep Peep from dashing inside. This was Peep's second day of being an "outside sheep" and he wasn't a bit happy about it. "Where are the boys?"

"Resetting their traps. They both had a fox this morning." I emptied potatoes, carrots, parsnips and turnips into the sink—there'd be stew for dinner tonight. "I gave them an hour lunch break."

"That must make about seven apiece for them," Milt said as he ran water over the sand-encrusted vegetables and started peeling carrots.

"**Seven for Randall** and six for Chuck," I laughed. "They would have been tied except for the one that got away."

"Yeah. Too bad about his boots, too. No way to put a decent patch right where the sole and upper meet. Has he tried them since I used 'Shoo Goo'?"

I nodded while I turned the browned chunks of meat in the Dutch oven on the back of the stove. "It still leaks."

"He'll have to stay out of the water, and up here that's almost impossible," Milt said as he dumped vegetables into the pot and I closed the lid. "He handled himself pretty well through that little incident, I thought." Milt dried his hands on his jeans and poked through the spice cabinet. "Are they friends again?"

"Ha!" I snorted. "When have they ever been friends? They're too competitive for that. I'm just glad they stop themselves short of actual combat." I shook my head. "Sometimes I worry about those two."

MILITARY SURPLUS. Not everything left behind by the military is useful. This old dock, for example, has seen better days. But buildings and docks in the background are used daily by the Holmes family.

Milt crumbled dried celery and parsley in his hand and added it to the pot. "They're good boys. Randy is just right at that age where he wants to push his limits."

The yard gate slammed. "Here they come," I said as I watched Randall dash down the walk and Chuck come around the corner of the barn. "Shall we have a holiday?"

Milt came to stand beside me and peered up at the clear blue sky. "Well, I'm going to check on the sheep in Peacock and West Point pastures this afternoon. I put the bucks out a month ago and I want to make sure they're staying with the flock." He looked at me with raised eyebrows. "Feel up to a jaunt?"

Not really. I felt like a nice hike up into the hills around Foggy Butte or an afternoon of beachcombing, and I knew any jaunt with Milt would not be leisurely. If I hadn't watched him and Randall wrestle all over the kitchen in the frenzied game of "scissors-paper-rock" night after night, I would have doubted he even knew how to play.

Still, I wanted an outing and if horseback riding wasn't my favorite sport, maybe I'd improve if I did it more. "Sure," I said.

CALL OF THE WILD. The sun may not shine every day, but when it does, the beauty of Unalaska Island is obvious. It's a wild and lonely place, but one that Cora, Milt and the boys obviously love and wouldn't trade for a more "civilized" spot.

"I'll get your horse then," Milt said.

The boys came in together, not openly hostile, but not cordial either. When I told them our plans, Randall immediately rushed after Milt. Chuck, however, wasn't so keen.

"Do you need me?" he asked.

"No," I said. "I just want out of the house."

"I have to write an English essay," Chuck explained. "I'll stay home and do it."

"Am I dreaming?" I asked. "Did you actually say you wanted to stay home and do class work on this glorious day, the only sunny day we're likely to get for another month?"

He gave me an offhand shrug. "It's for Gail."

Of all the instructors Chuck had, Gail Hocker was his favorite. She had made herself a real person for him, always sent personal notes with his lessons, encouraged him, praised his efforts and made clever criticisms about his atrocious grammar and punctuation.

I was grateful to her. Even though Chuck hated English and would much rather be reading his Alaska history course, which was all new and exciting, or better yet, his marine biology, which dealt with all the wonderful things that crawled and swam across the reef outside, he still did Gail's assignments first.

"That explains it," I joked. "Too bad Gail isn't your world history teacher."

On the way out the door, I pressed his shoulder, which was already hunched over a writing tablet. "Keep the fire going and don't let the supper burn." He mumbled something in reply.

When I reached the barn, Milt had my saddle on Stormy and Randall was tightening the cinch on our only mare, "Pixie".

"Where's George?" I asked.

"He's lame," Randall muttered. "Milt told me to ride Pixie. She's gentle."

He didn't look at me, just pulled himself up into the saddle and waited for us to go ahead of him.

No one said anything as we trotted single file along the narrow sheep track toward Peacock Point. The silence was not a comfortable one, and as soon as we crossed Sockeye Creek and the trail widened, Randall went ahead.

Something had happened. "Why is George lame?" I asked Milt.

"Because Randall has been running him on the beach to check traps before school." Milt sounded exasperated. "These horses have soft hooves from the wet tundra and shouldn't be taken on the rocks. Now George has a stone bruise."

I remembered taking Stormy around the beach the day Chuck and I had moved steers to the slaughterhouse. We had only done it because we were afraid we would lose them on the steep open hillsides. We had walked them slowly and carefully; still, we knew we shouldn't have been doing it.

"Does Randall know about not taking his horse on the beach rocks?" I was sure he had been told, but it must not have sunk in.

"He does now," Milt declared in a firm tone with enough emphasis that I doubted Randall would ever forget and run a horse on the beach again.

I sighed, feeling sympathy for both the forlorn figure on the horse ahead and for the stiff, erect back in front of me. I had known it wouldn't always be easy. Even love has its limits.

Suddenly Randall stood up in his stirrups and pointed. "Look, Milt," he called. "A cross fox."

A dark blur streaked across the tundra on a distant hillside.

"Good eye, Randall," Milt praised. "Even the dogs didn't see it."

The ice was broken. Never one to hold a grudge, Randall flashed a big grin and jog-trotted along the trail ahead. Milt, too, seemed relaxed in his saddle and the afternoon spread out before us with its joyous gift of sunshine and fresh air.

I even found myself enjoying Stormy's smooth trot. He was a big horse, for a Morgan; heavily muscled and strong. He had a gentle disposition, too, and I liked him.

I would have felt entirely comfortable on his back, except for the fact that he spooked at the slightest motion or sound that was the least bit different. Even a piece of plastic blowing along in the wind was enough to set him off. That knowledge kept me perpetually alert, hands clenched on the reins, prepared to ride out any sudden dashes or quick direction changes.

The sheep, with their shaggy wool, blended so well into the long brown grass they were difficult to spot, even large bunches of them. But the bucks had bright red brands on their backs and stood out like beacons.

For a couple of hours we rode through the pastures, gathering ewes together above Eagle Rock. When we finished, there were only 18 bucks. With his glasses, Milt discovered the remaining 22 perched on a craggy hillside. Randall and I kept the ewes bunched up while Milt took his dogs after them.

While we watched the flock, I noticed Pixie switching her tail and dancing sideways. "What's wrong, Randall?" He was on the opposite side of the sheep and Pixie kept pacing back and forth, whinnying and throwing her head.

"Nothing's wrong, Mom," Randall said. "She just wants to be over there with Stormy and I won't let her, that's all." He turned her in a tight circle so she faced away from me. But she still pawed and danced.

"Maybe you should get off and hold the reins until Milt gets back."

"She's fine, Mom. Stop worrying," Randall said with exaggerated patience. "Anyway, here comes Milt. Can I go help him?"

Without waiting for an answer, he let Pixie have her head and she galloped toward Milt and his little band of bucks.

"You be careful!" I hollered after him. Then I nosed Stormy around the sheep, who stayed huddled in a tight mass, putting myself in the spot Randall had vacated so when the bucks ran into the flock, I wouldn't be in the way.

By then, Randall had met Milt and put Pixie on the far side of the bucks, keeping them from breaking away down the hill. They came fast, and when the bucks caught the scent, they came even faster.

Bounding and leaping like lambs, they plowed into the flock. I nudged Stormy to one side and the whole bunch escaped past me, stringing out in a long line, back the way they had come. I watched them go, seeing the red backs mingled with the white, thinking about the new lambs we'd have in May.

As we started the 4-mile ride home, I relaxed. The whole afternoon had gone well and I basked in the knowledge that we had actually been some help, and I was really getting the hang of riding. The sun was warm on my back, we were headed to-

COMMON HORSE SENSE. Milt halters one of the ranch horses for another day of work. With free-ranging cattle and sheep, the only way to round them up is on horseback.

ward a warm house with supper on the stove. Randall and Milt were trotting just ahead of me in companionable harmony.

"Hey, Milt," Randall called, reining in Pixie. "Are those sheep on that hill?" He pointed toward the crest of the hill we were passing.

"I'll look with the glasses," said Milt, turning Fuzzy sideways and looping his reins over the saddle horn. Fuzzy stood motionless while Milt scanned the hills with his pocket binoculars.

While Milt studied the hills, I shifted in my saddle and looked across the 12-mile pass to Umnak Island. "Look, Randall, Mount Tulik is steaming."

A majestic white plume rose over the snowcapped volcano. In the still air, it drifted skyward hundreds of feet, distinguishable from the clouds by its very whiteness.

"Makushin is smoking, too," Randall said. "Look."

Sure enough. In the cloudless distance, we saw a narrow white pencil rising from the active volcano on our island, even though it was over 80 miles away on the eastern end, near the Unalaska village.

"There they are," Milt broke in, holding the binoculars to his

FIRE AND ICE. The snow-covered peak in the distance is Mount Tulik on Umnak Island. Mount Tulik is an active volcano that occasionally spews steam and ash. Its last significant eruption was in 1960, when several inches of ash fell on the ranch.

face with both hands. "Two, three, four, five…"

Suddenly, Pixie swung her hindquarters into Milt's horse and kicked out with both hind feet. Fuzzy stumbled and staggered to one side. Pixie kicked out twice more in quick succession, then whirled toward him, bucking and squealing, her mouth wide open.

Caught unaware, Milt lurched forward, then grabbed his reins and slapped the mare across the nose with them. "Whoa there, Pixie," he growled through clenched teeth. "Whoa, now."

It was over in an instant. Pixie dropped her head and stood quietly with her sides heaving. Randall still clutched the saddle horn, his face astonished. "What happened?"

The mare lifted her head and moved a few feet away, looked around her and began cropping dry grass.

"Pixie didn't think Fuzzy was paying her enough attention, I guess," Milt answered in a tight, strained tone, very unlike his normal voice. "Just being a mare."

Milt turned in his saddle and sucked in a quick breath. I saw the color go out of his face. "Too bad it was me she got."

"What?" My voice rose. "Oh no! Did she hit you?" I dismounted in a rush. "Where?"

"My ankle." Milt let out a slow, agonized sigh. "She smashed it into the stirrup."

"How bad?" I whispered.

"It hurts," he ground out.

"I better look." I stripped off my gloves and worked Milt's foot out of the stirrup, thankful he wore galoshes and loafers, rather than gum boots, which he hated and refused to wear. I would have had to cut one of them off.

After his foot was free of the stirrup, the rubber boot slipped off easily. Milt groaned deep in his throat.

"Sorry," I whispered. "Sorry."

His ankle was puffed out like a baseball. His sock was wet with blood. As much as I hated to touch his foot, I had to see what the damage was.

Gently I eased off his loafer and rolled down the sock to expose an ugly crescent wound just under the ankle bone. It seeped blood in a slow ooze. The skin around it was all dark red, angry looking, severely bruised and swelling rapidly.

"It could be worse," I told Milt, pulling the bandanna off my hair and wrapping it tightly around his ankle. "I can't tell if it's broken, but this will control the bleeding and the swelling."

I rolled the sock back up over the neat bandage and slid his boot over his foot. "Help me get your foot into the stirrup."

"AAAAHHHHH! OOOOOOKKKKKKAYYYYY!"

I supported his ankle with one hand and held the stirrup with the other while he maneuvered his toe through the opening. When we were finished and I looked up, sweat was beaded across his forehead.

"Think you can ride?" I asked.

"Easier than I can walk." He tried to grin, but it disappeared when his horse took a step. "We better get a move on."

I put his shoe in my saddlebag and mounted Stormy. Milt called to Randall, who was keeping Pixie away from us. "Chase those few in with the others, okay?" He grimaced. "And make sure everything stays across the creek. I don't want those bucks doubling back."

Randall nodded eagerly. "Sure, Milt. Don't you worry about the sheep. I'll take care of them."

"Watch that horse," I shouted. "If she acts up, get off and walk."

"Okay, Mom," Randall waved. "You just take care of Milt." He headed away from us, following the sheep.

Me? Take care of Milt? That would be a first! I urged Stormy up beside him. "We should go fast, if you can stand it," I told him. "We need to get that foot elevated and some ice on it." He tightened his lips and "click-clicked" Fuzzy into a canter.

Milt didn't complain once all the way home, even when we stumbled and crashed through creeks and gullies. He waited in patient agony while I struggled to open and close the heavy gates between the pastures. We rode right up to the door of the house and Milt dismounted in a clumsy jumble of arms and legs as I tried to help him. Balanced on his good leg, with an arm around my shoulders, he limped into the house.

Chuck met us at the door in horrified astonishment. "What happened?"

"I'll explain later," I panted. "Take care of the horses, please."

Milt hobbled into the bedroom and sank down on our bed. I helped him lift his injured foot onto the covers, pulling pillows under it until I had it elevated above the level of his head.

"I'll have to wash it," I said bluntly. "It will hurt." He closed his eyes.

I tried to be gentle, but I couldn't take a chance on any infection starting in the open wound. I washed it with Betadine and covered it with Neosporin ointment. The bleeding had stopped and my bandanna had halted the swelling. I put a Telfa dressing over the ointment and wrapped it in a snug Ace bandage.

"Now for some ice." From the storeroom freezer I took two packages of the frozen hamburger we had ground during butchering. I didn't have ice cubes. I wrapped the hamburger in towels and placed them around his foot.

Milt raised himself on one elbow and looked at the bulky bandage. "Is it broken?" he asked.

"No way to tell without an X-ray," I replied, shaking two codeine tablets into his hand and handing him a glass of water. "She hit you right on the ankle bone, so she missed both the long bones in your lower leg. If anything, the blow may have chipped the ankle bone."

I watched him swallow the tablets. "I'm doing everything here that a physician's assistant could do in the village, and that is rest, elevation, ice and compression." I shrugged one shoulder. "What do you think?"

He handed me the glass and sank back onto the pillow with a grimace. "I think it's a long way from my heart." His eyes twinkled and he even managed a slight grin. "Cover me up, will you? I feel sleepy."

Both boys were in the kitchen when I finished. "Is he all right?" Their faces turned to me anxiously.

"Of course, he's all right," I assured them. "Just a bad bruise and he'll need to stay off it for a few days."

Randall gave a relieved sigh. But Chuck looked at me for a long heartbeat, his eyes full of concern. But all he said was, "I'll get the lights."

STRENGTH IN FAITH. With appropriately descriptive Bible verses always on view, sometimes, as when the horse smashed Milt's ankle, there was only so much that could be done and the rest was left to prayer.

"Oh, yeah," Randall jumped up. "And I'll feed the horses and put up the dogs."

"Milt will appreciate it," I thanked them.

After they had gone, I watched darkness fall through the kitchen window and thought about all the things I had left unsaid. Shattered metatarsals, fat embolus, bone infections, blood clots, sepsis.

Like a great topographic map, our location and that of the nearest medical help stamped itself on my mind. Eighty air miles to the village clinic…800 air miles to Anchorage and the nearest hospital.

It was dark and it was winter and we were all alone.

I closed my eyes and prayed.

Chapter Thirteen

ONE GOOD SPLASH
DESERVES ANOTHER

For 24 hours, Milt's foot hurt enough that he stayed in bed. But by the second morning, he unearthed the crutches from his hip surgery and hobbled out to the barn.

After 10 days, he discarded the crutches. "I'm well," he announced on Groundhog Day, sliding his foot under the breakfast table with a slight wince.

Afterward, I changed the dressing for the first time. The swelling was gone and the red-purple bruise covering his entire foot had changed to a muddy yellow-black. Peeling back the Telfa, I inspected the healing wound edges, bending my head close enough to smell.

No infection.

"Looks good," I answered the question in his eyes. The damaged skin had sloughed off, leaving a healthy pink tissue underneath. "Not crooked either."

Randall sucked in a deep breath. "Man, she really whacked you." He leaned on the arm of Milt's chair, watching me pour hydrogen peroxide over the wound. It foamed and bubbled.

"Ouch!" Chuck squirmed and turned away his face. "I bet that hurts."

"Not much," Milt gritted.

"I bet it's almost as bad as when I cut my hand on my pocketknife," Randall bragged. He showed Milt the wide scar between his thumb and forefinger. "I almost bled to death."

"Huh!" Chuck scoffed. "That wasn't as bad as the time I cut that galvanized tank with an acetylene torch and burned my lungs. My chest hurt so bad I thought I was dying."

"That's nothing," Randall jeered. "How about when I cut my knee on the tin? Mom even wanted to put stitches in that." He folded his arms across his chest and stuck out his lip. "I bet that was the worst accident that ever happened here."

"It wasn't as bad as the time I had that fish hook…"

"How about the time Fuzzy kicked Mama?" Milt broke in.

Randall looked at me in surprise. "How should I know?" he grumbled. "She never let me see the bruise."

"Trust me," I frowned. "There was a bruise."

"I believe you," Chuck volunteered. "I carried you into the house, remember?" He brushed a hand across his eyes. "I saw the hoofprint."

"Just a bad bruise," I said. "And my own fault, too." I re-wrapped Milt's foot and replaced his sock. "Served me right for being careless."

I hadn't thought about that day in a long time. It wasn't a good memory. I had walked up on the wrong side of Fuzzy and he kicked me squarely in the abdomen. No one could have made me believe the incredible power in a horse's hind legs until I experienced it firsthand.

At first I had worried about a ruptured bladder, then later, when I tried to walk, about a broken pelvis. I had been lucky. Except now I wasn't really comfortable with horses anymore; didn't trust them and didn't have much faith in myself around them.

"Okay, Milt." I slid a slipper over his sock. "Why not rest it awhile. It's pouring rain anyway."

"Oh, good!" Randall exclaimed, pulling a chair close. "Let's play some more 'scissors-paper-rock'."

"Not you, chum." I shook my head. "You don't have a sore foot."

"Aw, please?"

"What?" I expected begging from Randall but not Milt. "He has to study," I protested.

"One game," Milt coaxed. "One quick game."

It was the most bizarre contest I had ever seen, and they loved it. Scissors cut paper; paper wrapped rock; and rock broke scissors. They had hand signals for each object and when they had counted to three, each gave a hand signal.

The one signaling the object that did the most damage won and got to hit his opponent on the wrist with two fingers as hard as he wanted.

"Dumb game." Chuck observed as Milt whacked Randall smartly on the wrist, while Randall closed his eyes and tried not to flinch.

I agreed, watching them smack their fists into their palms as they counted.

"Rock!" Randall shouted. "You did scissors. Rock breaks scissors. I get to hit you."

Milt held out his wrist. Yes, it was a very dumb game. I sat down at the table across from Chuck and turned up the lamp wick.

"How is the essay coming? Did you finish it yesterday?" He had his books opened at his side on the kitchen table.

"Almost. I'm recopying it now." He covered the paper with his hand.

"May I read it?"

"Not till I get it back." He twirled his pencil. "You're too strict. I'd never get it perfect enough to satisfy you."

"I know what you're capable of," I countered, "so I expect more."

"You expect the impossible," he grumbled, copying the last lines and wadding his original into a tight ball, which he threw at me. I swatted it back and we played Ping-Pong across the table until Randall's shriek interrupted us.

"I beat! Two outta three! I beat!" He ran to the sink and splashed cold water on his reddened wrist. "One more game, Mom, please?" he begged. "Just one more?"

"No."

"Aw, please?"

"You heard Mama," Milt said. "Time for school." He stood up and reached for his coat. "I'll be in the barn."

While he got into his rain gear, I fixed a warm bottle. "Feed Peep, please." I handed it to him.

"Whose pet is he?" He raised his eyebrows.

"Ours," I teased, kissing him quickly. "Baby your ankle now."

With a big sigh, Randall dragged out his books. As I sat down beside him, he whispered "Attila" under his breath.

"I heard that," I warned.

"You're so mean." He glared at his book. "Meaner than Attila, I bet."

"Someday you'll thank me," I retorted. From the mutinous look on his face, I could tell it wasn't going to be soon. After quizzing him on his weekly spelling list, I listened to him read and helped with the first two problems of his math assignment.

After I finished, I saw Chuck was deep in his marine biology textbook, so I blew out the lamp and sat down at my spinning wheel in the corner of the kitchen.

Up and down. Up and down. I treadled the machine with a soft whir. Since the day I had changed the tension and discovered such a difference, I had practiced every spare minute. Yesterday I added something new.

The book suggested spinning store-bought yarn until I got the feel of treadling and feeding the flyer at the same time. I felt stupid spinning yarn that was already spun. But it worked!

For the third time I threaded the end of a ball of yarn I had unraveled from an old sweater through the flyer opening and tied it to the bobbin.

Up and down. Up and down. The yarn slid through my fingers and wound on the bobbin. The fact that I was making the wheel work gave me such delight I could hardly wait to experiment with actual wool.

Last night, in anticipation of this success, I washed some wool. First I soaked it 10 minutes in 180° water and dishwashing detergent. Careful not to felt the wet wool by twisting the fibers together, I had squeezed out the excess water and now it was drying on the glove rack above the stove. Every hour I turned and fluffed it. It was beautifully white, and so soft it was hard to believe it came off a range sheep.

"Hey, it's time for lunch." Randall slammed his book shut. "Are there any leftovers from last night?"

I let go of the yarn end and it disappeared through the orifice. With a satisfied sigh, I looked at the clock.

"A couple of lamb chops, I think." I stood up, stretching and rubbing my neck. "Look on the porch."

In a minute I heard him rummaging in the cupboard that

served as our winter refrigerator.

A sheet of paper came zinging across the table at me.

"There. You wanted to read it so bad, go ahead." Chuck closed his books and stacked them together with a half-defiant look.

It was his essay.

"What made you change your mind?" I asked.

He shrugged. "I want to know what you think—about commas and stuff."

He fidgeted for a minute, then made himself a peanut butter sandwich with leftover sourdough pancakes.

I read the neat lines, looking for anything that would lower his grade. By the time I reached the second line, I was immersed in his story. It was delightful—witty and well-told.

It wasn't until I read it a second time that I noticed his almost complete absence of capital letters, paragraph indentations and punctuation marks.

THE HALIBUT
By
Chuck Harvey

When we came to Alaska last year we hadn't ever caught a halibut. So when we were in dutch harbor we went to visit a friend. he invited us to go fishing with him after he got off work.

that night Randy and I got into Jim's sixteen foot seine skiff and took off. randy and I had our fishing pole and Jim had a halibut jig with the line wrapped around an old bone. Randy and I caught pogies and black bass but jim wasn't catching a thing with his outfit. I told him it was bad luck to have his line on a bone. so he gave randy the jig and he took the pole. Ten minutes later randy had an enormous bite. He just hung on to the line and the skiff was doing a slow circle so Jim said it must be a halibut. randy couldn't pull the line up by himself so I had to help him, by the time we got him to the surface we were both dead tired. Jim said we'd better shoot him, so out comes this gun of some monsterous caliber, the first shot was a dud, and I think the fish heard the click because he started to fight a lot more. the second shot went off and killed him along with the cheeks. it took all three of us to get the slimy thing in the boat. he flopped all the way back to town. the thing was almost 4 feet long and weighed about 80 pounds. randy only weighs 90 pounds so it's a good thing he wasn't out fishing alone

"Well?" Chuck asked me.

I looked at the careful margins, the neatest handwriting I had ever seen on an assignment of his, and still I itched to correct the one misspelled word and put in all the capitals he had missed. His earlier comment about everything having to be just perfect before I was satisfied still lingered, and I resisted. **"It's great,** Chuck." I meant it. "You tell the best stories. I know Gail will like it."

He looked startled. "You mean I don't have to do it over?"

I shook my head. "Send it in the way it is. A few errors aren't going to knock down your grade that much." Let Gail give him the bad news. He would take it better from her.

"You really liked it then?" His eyes watched me; expectant, unsure.

"I think it's terrific." The look in his face unsettled me. Attila? Perfectionist? Never satisfied? Strict? Was that what I had become?

I brushed the thought aside and handed him the paper. "What ever happened to that fish?"

"We left it at Jim's," Chuck said. "We didn't have room on the plane."

"That's right," I remembered. "Milt and I had just gotten married."

Jim Dickson was the first friend we had met after coming to the ranch. He was the skipper of the crab boat *Shellfish* and an avid sportsman. During our first year here, before I married Milt,

FOXES AND FISHES. All that reading Randall did in his trapping book obviously paid off. He's a pretty good fisherman too, it seems. Randall smoked these salmon after soaking them in soy sauce and brown sugar.

Jim had anchored out in the bay and kayaked ashore. That was the beginning of a deep friendship. The boys admired Jim and looked forward to his visits. He came whenever his boat was in our vicinity and he always brought our mail.

"Can I have these potatoes, too?" Randall called.

"No, they're for supper." I stuck my head around the door. "Eat bread and peanut butter if the chops aren't enough."

I was teaching the boys how to cook, but neither of them were much interested. Chuck's only accomplishment was boiled rice, and Randall preferred to eat food cold if I didn't cook it for him. He would put peanut butter on a piece of bread and heat it up in the oven, though, if he couldn't find leftovers.

While the boys ate, I made a cup of hot chocolate. Pouring

it in a lidded cup, I put on rain gear and splashed through the downpour. Milt did not break for lunch. Sometimes I got him to drink the hot chocolate; other times I went back hours later and it was cold and untouched.

I found him in the barn, repairing a wooden pulley. I gave him the cup and wiped my streaming face. "How's your foot."

"Okay." He gulped the warm cocoa. "Thanks, but you shouldn't have come out in this rain."

"It feels good." I listened to the steady drumming on the tin roof. "And so warm. It must be 40°."

"That reminds me." He bent over the anvil. "I looked at the last pelt the boys got and it was ragged from rubbing. It's so warm, they're getting rid of their winter coat."

"Time to quit trapping, then?" I asked.

"I think so." Milt hammered the metal he was shaping. "They'll find mates and den up soon."

"Good," I said. "I hate the smell and the mess. I'll never get that musk odor out of Randall's jeans. I think he rolls in it."

Back inside, the afternoon classes dragged. After I read a chapter from *Hans Brinker and The Silver Skates*, they opened their books with weary resignation.

The boys didn't like me to hover and coach while they studied. When they needed help, they asked for it. To keep my hands occupied and my mind free for questions, I usually practiced spinning or sewed.

The last few afternoons I had been making curtains for the kitchen windows, which seemed naked and cold without them. I had found a box of unused sheep carcass muslin in the attic. Used to keep meat clean after the sheep had been dressed out, they were 40 inches wide and 60 inches long, and ecru colored.

The treadle sewing machine I used had belonged to Milt's grandmother. It was a Singer and, next to my spinning wheel, my most treasured possession. Unlike the wheel, the sewing machine worked perfectly and even had attachments like a button-holer and ruffler.

But the curtains I was making were simple Priscillas. Without any pattern, I couldn't get fancy. The Christmas box from Mom and Doris had contained 12 packages of rickrack in dif-

STITCH IN TIME. There's no tailor nearby, so when things get torn, Cora takes to the old treadle sewing machine and makes the needed repairs. While each member of the family depends on the others, each also has to be pretty much self-reliant.

ferent colors, so I planned to decorate the simple flounces with them.

At 3, the boys put away their books and got out their art projects. Chuck had taken an elective course on tanning and taxidermy and had tanned a strip of sheep hide in battery acid. Now he was working to make it soft and pliable. Randall was weaving a rya wall hanging on the rigid heddle loom.

Since this wasn't really class work, we talked whenever we wanted.

"Tonight's Saturday night," Chuck commented. "Guess I'll go out carousing and drinking root beer." He looked out the window at the pouring rain. "Maybe I'll call Sally or Jane or Cynthia. One of them might like to go carousing, too." He blew on the pane and rubbed it with his shirt sleeve.

"Ha, you dreamer." Randall snorted. "You'll never get girls to stay out here." He wrapped yarn around his shuttle. "All I want is a horse, a gun, a dog and a boat, and I can stay out here forever."

"I like those things," Chuck countered. "But I like girls, too."

UNALASKA WIND CHIME. If you want to get the gang in for supper, it takes a mighty big bell to be heard over the wind and the surf at ranch headquarters. This old red clanger looks like it can get the job done.

"What you need, then," I quipped, "is a girl who likes horses, dogs, guns and boats."

"And where will I find her?" Chuck's face twisted derisively. "Behind a tree? Get real, Mom."

"It only takes one," I said mildly. "Milt found one."

"You?" Chuck laughed explosively, grabbed the curtain I was working on and shook it under my nose. "You like picking flowers, reading romances, walking hand in hand on the beach and..." he shook the curtain again, "ruffles."

"Yeah," Randall interrupted. "And you hate guns and horses. You're scared of the boat and you won't let the dogs in the house."

"But I love Milton," I defended. "That's what's important."

"Yeah." Randall looked up from his yarn. "I'm never getting married till I find a girl who loves me as much as you love Milt."

"Good luck," Chuck snorted, and returned to his leather.

Folding the unfinished curtain, I reminded him of an earlier discussion.

"You know you can go Outside for school, if you want to," I said. "Our only requirement is that you get decent grades. You can't just goof off."

"Nah." Chuck contemplated the leather between his hands as he moved it back and forth across the edge of the table. He looked up and grinned. "I don't like girls that much."

"Four o'clock, Mom," Randall interrupted. "School's out."

After the door slammed, I watched them splash down the walk. Randall jumped in every big puddle. There was a small lake growing beside the chicken house. When he jumped in it, water flew above his head.

I sighed. Thank God for rain gear and gum boots. I turned away from the window and, after a stern talk with myself, started cooking instead of carding a rolag of the wool, which I knew from frequent turning, was now dry. After supper, I promised myself.

Cooking for Milt and the boys was easy. They wanted meat and potatoes and vegetables. And they didn't want it all mixed up—no tuna casseroles for them. Sometimes I got away with macaroni and cheese, but not often.

So I fried lamb chops, hash-browned the cold potatoes,

splurged on two cans of hoarded asparagus—everyone's favorite—brought in a bowl of green Jell-O with pears in it for dessert, sliced a loaf of bread and rang the dinner bell.

When the last pear disappeared, Milt pushed back his chair. "Chore time." He retrieved his coat and hat from behind the stove. "The rain has stopped."

I noticed his limp was more pronounced than it had been at breakfast.

"There's still enough light for you boys to pull up your traps, if Mama lets you off the hook with dishes," he said.

I nodded and they both vanished before I could change my mind. In a few minutes the generator hummed to life, full coal buckets thumped on the back porch and chunks of wood hit the wood box.

It was nearly dark outside and I was just sitting down with my new wool at the table when footsteps sounded on the porch and the kitchen door opened.

I turned around. Chuck stood there with the strangest expression on his face. He looked expectantly anxious and devilishly gleeful at the same time.

"Randall's really in for it now," he said in a hushed voice. "He splashed Milt."

Oh, no. A picture of him jumping in that puddle with the water shooting skyward flashed across my mind. "What happened?"

"Well," Chuck giggled nervously, "Milt splashed him first; you know how they always scuffle back and forth. But it was just a little splash." He glanced backward through the porthole in the kitchen door and lowered his voice. "So Randall splashed him back."

"That sounds fair enough," I reasoned.

"Yeah, but Randall wouldn't stop. He kept splashing Milt in every puddle they came to. Finally Milt told him to knock it off and if he didn't, the next time he splashed him, Milt was going to throw him in the big one by the chicken house." Chuck gasped a big breath.

"So Randall took off, but when Milt went around the chicken house, there he was, waiting, and he jumped in that big puddle and splashed water all over Milt," Chuck's face was bright red. He shook his head and stared at me with astonished disbelief. "I still can't believe he did it."

"Where's Randall now?" I hissed between my teeth.

"Somewhere out in the horse pasture," Chuck said, going to the window and peering out. "He ran and Milt couldn't chase him with his sore foot."

"Then where's Milt?"

"Sitting on the porch, waiting for Randall." Chuck giggled again. "Randall came back once, but saw Milt and took off again."

I grabbed my coat and headed for the porch. Milt sat on a stool beside the door, waiting, water dripping off his clothes in tiny rivulets, splashing on the floor. "I'll go get him," I ground out.

"I wish you wouldn't," he said gently. "This is between Randall and me."

"What are you going to do?" I stabbed my feet into my boots and jerked on a slicker.

"Throw him in that puddle, like I said I would."

I glared at him; mad at everybody now. "He could stay out all night."

Milt's green eyes glittered like ice in his stern face, but his voice was whisper soft. "And I'll be here waiting."

I closed my mouth in a thin line and stomped past him. Outside in the dark, I ran into Randall at the corner of the house. I grabbed for him but he dodged my hand and danced away.

"Get in the house, right now!" I commanded in a harsh whisper.

"Is Milt on the porch?" His voice sounded tiny and scared.

"Of course, he's on the porch." I walked slowly toward his voice, seeing his shape as a blur through the mist. "He'll be there till the second coming."

"Then I'm not coming in."

"Don't be foolish," I said. "You might as well come in and face the music."

"No!" he shouted. "Keep away from me!" His footsteps were a series of staccato splashes as he raced away.

I followed, stumbling and slogging through ankle-deep water in the horse pasture. "Come on, son. You know this is ridiculous."

No answer.

Twice I thought I saw him as a dark flash ahead of me, but I knew catching him was hopeless. I could never match his speed, much less overtake him. Chasing after George had made him fleet as a deer.

"All right," I hollered in defeat. "But if you catch pneumonia, don't expect me to feel sorry for you."

Turning around, I started for the house. Furious, worried and entirely at a loss for a solution. My footsteps slowed and I caught myself splashing angrily in every pool I stepped in. Before I knew it, I was jumping up and down angrily in 6 inches of water.

Suddenly Randall's slight frame shot past.

My anger disappeared as relief flooded through me. I waded out of the puddle and retraced my steps, but I didn't hurry. In fact, I detoured past the house and walked out to the barn and played with Peep awhile, fed him some ground-up oats we were teaching him to eat and petted all the dogs.

The first thing I saw when I did reach the house was a pile of sodden clothes right outside the back door. I recognized Randall's sweatshirt and jacket. I girded myself for the

IN A RUT. This is what passes for a road on the ranch. It's not bad, considering that there are few roads at all and horses provide most of the transportation.

172

battle and stepped inside.

There in front of the stove, with wet tousled hair and wearing clean, dry long johns, sat Milt and Randall. Facing each other on chairs, they ignored me.

Slap! Slap! Slap! Fists pounded palms.

"Aha!" Milt exclaimed. "Scissors cuts paper. Put out your wrist."

CANNED PUZZLES. Supplies are ordered for 2 years at a time and arrive by boat. When one load got soaked, some of the labels came off. One morning, Milt grabbed a can of what he thought was milk, punched two holes in it and tried to pour it on his cereal. He was surprised to find the can held creamed corn! Since then, Cora has learned to read the code numbers on the bottoms of the cans and Milt no longer has to worry about adding corn to his cornflakes.

Chapter Fourteen

THE MALL AT CHERNOFSKI

March came in like a lamb, but by the end of the first week, typical Aleutian weather returned and we were driven inside by raw winds and bitter cold.

But that tantalizing glimpse of spring was enough to send me to the "shopping center" on a stormy Saturday afternoon. Everyone else was already there.

I rummaged through the big wicker basket that dominated one corner of our kitchen, pushing aside catalogs from Sears, Eddie Bauer, Bloomingdales and Mary Maxim's, until I found last year's seed catalogs. Spreading them on the table, opposite Chuck's and Randall's stacks, I sat down next to Milt.

He looked up from his Nasco West. "Going to try some gardening?"

I opened the first pamphlet and pored over the pictures of bright-orange carrots and dark-green spinach.

"I know it's futile," I sighed. "Last year the potato starts rotted in the post office. They were glue when they finally got here."

I turned the page and visualized a salad with tiny radishes and cherry tomatoes.

"And the year before that the slugs ate every potato plant that survived the wind and rain long enough to get 3 inches high." I cupped my chin in my hand and sigh again. "I need a greenhouse."

Milt gave me a dubious look. "Do we get enough summer sun?"

"We could grow leaf lettuce and spinach, cool-weather crops, maybe even peas and potatoes." I turned the page. Large luscious strawberries stared at me. "I want to try."

"Well," he scratched his chin, "let's see. First we have to go around to all the pasture fences; then we have to go to Wood

Cove and chop up that cedar log before a storm takes it off the beach again, then we have to get coal." He picked up a seed catalog and leafed through it. "By then it'll be time to start rounding up sheep."

"I'll help build a greenhouse," Chuck offered, looking up from the stamp he was peeling off his addressed envelope.

"Me, too," Randall chimed in. "I like gardens. I had a great one in Idaho."

"It wouldn't be too much of a job, Milt," I said. "If we built it on the southeast end of the old laundry house and used those windows you salvaged from the military-base Quonset huts."

Milt grinned. "You have this all thought out, don't you?"

"I've thought about it a lot," I admitted. "I really want one."

"Okay, we'll do it." He dropped the catalog back on the table. "The biggest job will be finding the lumber."

"The whole back of Cutter Point warehouse blew out in the past storm," Randall said. "I had traps over there and they al-

LUMBER LITTER. *What time starts, the weather on Unalaska Island usually finishes. Some of the old military buildings on the ranch have seen one too many storm and end up so much material for another building project. Nothing like having your own lumberyard.*

most got clobbered."

"Then that's the first place we'll look," Milt told him.

"What are you doing, Chuck?" I asked.

He looked up with an exasperated scowl. "I'm trying to get this blasted stamp off without tearing it so I can use it on another envelope."

"Why?"

"Because 3 weeks ago I wanted these sailing dinghy plans, but I changed my mind." He scratched at the ragged stamp. "I found a wrist rocket I wanted more."

"What do you want with one of those?" I took the envelope and held it over the teakettle until the glue softened and the stamp slipped off.

"Target practice," he replied. "And to move cattle along with, and maybe get a ptarmigan, if I'm lucky."

"That'd take some luck," Milt put in. "I'll make you a slingshot like I had when I was a kid. You can put a rock anywhere you want with it."

"Great!" Chuck reached for the Cabela's catalog. "I'll have enough for a T-shirt." He wrote furiously. "Okay, I'm finished." He dropped his letter in the mail basket. "Can we make the slingshot now?"

I put the stamp on the blank envelope and stored it in the desk, thinking about how getting mail six times a year sure cut down on impulse spending.

"How about you, Randall?" I asked. "Did you change your mind?"

"Nope." He scribbled his return address on the envelope corner. "I'm getting logwood dye for my traps and wax to keep them from rusting." He handed me the order blank. "Do I have this much money?"

I kept a ledger book of the boys' earnings and whenever they wanted money, I wrote a check. We didn't have much access to cash.

"I think you have enough." I opened the checkbook.

"If I don't, can I charge it?" he asked. "I'll make more money when we round up and shear the sheep."

"Cash on the barrelhead," Milt called from across the room

where he and Chuck were busy measuring rawhide thongs. "Two things a man should never do," he said. "One is buy a dead horse and the other is pay a kid before the job is done."

I laughed. "I'm afraid I agree with you there." I wrote Randall's check and handed it to him. "But you have plenty and some left over." I showed him the ledger book.

"Enough for an Archie comic book subscription?"

"Yes. Why?"

"I like something in the mail when it comes that's just for me."

I looked at his earnest face, sorry that I hadn't considered it before. I always read letters aloud whenever they came from grandmothers and aunts. They were addressed to all of us but I was the one who opened them.

Mail was a serious pleasure for Randall. He looked forward to the infrequent mail days with keen anticipation, while I almost dreaded the news that came; the terse reminders of overdue bills and lapsed premiums.

"That's a great idea," I told him. "Go ahead and order one."

While he filled out the subscription, I glanced at the letter

BURNER AND BYTER. Just because an old-fashioned stove heats the room doesn't mean one has to write with a goose-quill pen. The word processor has helped Cora crank out hundreds of letters—and this book.

box, noticing Milt had added several more. "Why this flurry of letter writing?" I asked him.

"Oh, didn't you hear? You must have been out in the storeroom." Milt looked up from the knots he was tying in the thongs. "Scott on the *Valiant* called. He'll be here next week to store pots before he goes to Dutch Harbor. He offered to take our mail to the post office."

"Then I better finish the grocery order right now. That's what I was doing in the storeroom, making a final check on what needed replenishing." I scooped the seed catalogs back into the basket and pulled the inch-thick Acme Provisioner from its cubbyhole in the desk.

Acme Provisioners was a ship supplier located in Seattle, Washington. They dealt in large quantities; 40 cases being their minimum order. They were efficient, quick and almost like family because Milton had ordered from them for so many years. I opened the catalog and began marking items where I had left off earlier.

Because our location made us so hard to reach, we ordered supplies only once every 2 years. Although there was now an Alaska commercial store in Dutch Harbor and a general store in Unalaska village, both were 80 miles away and might as well have been on the moon, as difficult and expensive as it was to get supplies from either place.

Those stores faced the same difficulties we did, as their suppliers were located in the lower 48. The produce they ordered was subject to the same rough weather and lengthy delays. Customers were loathe to buy wilted lettuce, moldy cabbage or apples that looked like they'd been rolled up the Alaska Highway with a stick. So, any produce that survived the air or sea voyage in reasonably good condition had an extremely high value.

To add to the cost, anything we purchased from the village had to be brought out by charted plane or boat. No commercial carriers had a scheduled stop at Chernofski since the mail plane was discontinued in 1977.

"When is the *Silver Clipper* coming out of Seattle this year?" I asked Milt.

"Sometime in the fall. We have plenty of time. Acme only asks for a 10-day notice."

"We're almost out of potatoes now. Couldn't we get our groceries on the *North Star III*, the boat that picks up the wool?"

"We could, yes, but they go past here in May on their way to Point Barrow, so potatoes and apples would come from last year's crop." Milt cut a neat leather rectangle and punched holes in each end. "Besides, they stop at small villages all along the

PIE, OH MY. Case lots of groceries like flour and sugar are converted into good things to eat, by Cora. Hard work in rough weather makes for big appetites that require hearty food. After all, it's not like the Holmeses don't burn off the calories.

coast, where Mike comes across the Gulf in 10 days with potatoes right out of the ground."

"You hate instant potatoes," I argued.

"But I hate mushy apples worse."

"You win," I conceded with a laugh. "Mushy apples are worse than instant potatoes."

I turned pages until I came to the "deli" section where I looked at the selection of waxed cheese. Just thinking about a crisp red apple conjured up images of Gouda cheese cut in thick wedges.

"Can I have Pop Tarts?" Randall scooted his chair around the table to sit beside me.

"Yeah," Chuck agreed, pulling out a chair on the other side. "And get more real milk chocolate chips."

"You mean like the ones hidden upstairs under the sofa and cornstarch?" I lifted my eyebrows.

"No, those were mint-flavored. I want the kind you hide behind the No. 10 cans of garlic salt."

"Chuck, you're impossible," I grumbled. "I had close to half a case, last time I looked."

"Think of all the cookies I saved you from baking," he reasoned with a cherubic smile. "And you didn't tell me I couldn't eat them; only the M&Ms and Milt's lemon drops."

I heard Milt chuckle as I checked off another case of chocolate chips and turned to Randall. "Do you have anything to confess?"

His face looked as angelic and innocent as Chuck's. "I might have chewed a couple sticks of your gum."

"My super bubblicious, double trouble bubble gum?" I shouted, rushing to the bottom drawer of my treadle sewing machine.

"I didn't take it all," he hastened to explain.

"Well, you might as well have." I tossed the remaining chunk across the table at him.

"Thanks." He calmly unwrapped and popped it into his mouth. "You said it made your jaws ache anyway." He blew a bubble big enough to cover his face. "Besides, you ate my last licorice rope."

"Oh, that's right, I forgot." A guilty flush tinged my cheeks. "Now we're even."

"Sounds like you guys need new hiding places," Milt said as he came up behind my chair and looked over my shoulder. "Did you order green tea?"

"Six cases," I said. "At least you won't have to worry about the boys drinking that."

For the next 2 hours we discussed, checked, deleted and let our mouths water. By evening we had chosen 67 cases of staples: 600 pounds of flour, 200 pounds of white sugar, 100 pounds of brown sugar and 100 pounds of assorted beans and rice.

We all watched as Milt added up the long column of figures

on his adding machine. It seemed like a long time before he hit the total button.

Tearing off the tape with a flourish, he announced, "$2,800."

I gulped, remembering the times I'd been down to my last dollar, choosing between eggs and bread. It sounded like such a huge amount.

"Not bad for 2 years," Milt went on. "Our beef sales will cover it and pay for fuel, too." He looked over at me. "This is less than last time, isn't it?"

"Maybe I'm getting the hang of it," I smiled with a hint of mischief. "Or maybe I forgot something." I reached for the catalog.

"Oh no, you don't," Milt laughed and caught my hand. "Time for supper. Grocery shopping always makes me hungry…Uh, oh. Hang on."

We all heard it—a distant roar, muted by the wind but somehow still distinct and all the more ominous because of the rain lashing the windows.

"Earthquake," I whispered.

We looked at each other, eyes anxious, and braced ourselves, and waited.

It came like a slow wave; a ground swell that lifted the rear of the house first. We heard bottles hitting the floor in the bathroom, then the kitchen range shuddered, making the tea-kettle and stove poker bounce and rattle against the metal.

We sat uneasily at the table as the worn kitchen linoleum lifted and settled as from a strong wind gust. Ever so gently, we felt the motion under our feet. The hanging plants in the big window swayed and bumped into the glass. Then it was gone.

"Ahhhh." Randall's relieved sigh voiced all our thoughts.

Not the big one. Not this time.

I broke the silence. "Just a tremor," I said cheerfully. "Nothing to worry about." But my voice wasn't quite steady, and when I opened the stove to add more coal, my hands were shaking.

"It's weird how they always start in the bedrooms," Chuck said, the color returning to his face. "And no matter how many we have, I can't get used to them."

"That side of the house faces south," Milt told him. "The

epicenter was somewhere in the Pacific Ocean, most likely the Aleutian Trench. The Pacific 'Rim of Fire' goes along the islands on that side."

Chuck swallowed. "Will we ever get a big one?"

"We could," Milt pushed his chair back and stood up.

"What would happen to us?" Randall came to the stove and stood very close to me.

"Not much." Milt grinned at him. "No trees or skyscrapers to fall on us here."

"How about a tidal wave?" Chuck cupped his hands around his eyes and peered out the window at the bay.

"Not here on the Bering Sea side." Milt reached behind the stove for his chore coat. "But the Pacific side has had some colossal ones. All those logs we see a half mile inland when we're riding came from tidal waves. We had one once that wiped out all the line cabins on that side of the ranch." His eyes sobered. "If you're over there and feel an earthquake, get the heck away from the beach."

"Hah!" Chuck exclaimed. "You don't have to tell me that." He turned back to the room. "I bet I could outrun my new slingshot." He picked up the simple weapon and swung it around his head.

"Thanks, Milt." He stuffed it in his pocket. "Want to go help me try it out, Randall?"

"Not till Milt makes me one," Randall replied. "I'm not even going outside until I find my football helmet."

"You chicken," Chuck said, grabbing his coat and following Milt.

After the door closed, I checked on the lamb roast I had in the oven and put a pot of rice on to simmer. Getting out the plates, I started setting the table. Randall brought the glasses and silverware.

"I appreciate your help, son, but you better get your coal hauled in before it gets dark."

"Mom." He clattered a spoon beside each dish. "I have a bump."

"What did you do," I asked absently, "fall down or run into something?"

"Nah, it just came."

My hand stilled. I looked into his puzzled face. He caught my eye, then looked away, tapping a fork against the table with rapid, nervous bangs.

"Where is it?" I kept my tone casual.

He took my hand and guided it underneath his sweatshirt. "Right here."

Under my fingers I felt it, a firm lump, smooth and round and movable, less than an inch from his left nipple.

Chapter Fifteen

A MOTHER FIRST

When the *Valiant* left the harbor a week later, Randall and I were on it. The lump on his chest hadn't changed. It didn't hurt. It wasn't inflamed. It was simply there.

I told him it was nothing to worry about. He would soon be 13, his voice was getting deeper and he had grown 2 inches since Christmas. I was *almost* certain the growth was a normal sign of growing up and changing. Almost.

I had a good medical education. I had worked in both emergency rooms and surgeries before settling on neonatal intensive care as my specialty. Before moving out of the mainstream of civilization, I had figured that my experience would come in handy in dealing with medical emergencies on the island—cuts, falls, even broken bones. Those I felt capable of dealing with, and had done so without a second thought.

But I hadn't reckoned on feeling so uncertain about something like this. All week I had pored over my medical books, looking for a statement that told me lumps like the one Randall had were a normal part of adolescence. I didn't find any and I had seen too many bald, scarecrow children to ever take any lump lightly.

When it came down to making a diagnosis, I couldn't take the responsibility, not for my own son. He trusted me and that frightened me more than anything—his life could depend on my judgment.

Randall wasn't worried. After showing me the lump, he forgot about it. Even when I told him we were going to see a doctor, he just shrugged and said, "Okay". I envied him his explicit faith and wished I had even half his confidence.

I also wished I knew more about the village clinic. Neither Milt nor I had ever been there. We didn't even know if it had a doctor. We knew from listening to ships with injured crew members that there was a physician's assistant and a registered nurse. If either of them thought Randall needed to see a real doctor, then we would have to fly to Anchorage, 800 miles away. I didn't want to do that. Anchorage was the largest city in Alaska; nearly half of the state's half million people lived there. Just the thought frightened me.

I looked at him standing beside me in the *Valiant's* wheelhouse, clutching the window ledge, his eyes glued to the horizon, trying to ride with the slow, lazy roll as the boat dipped its bow into the calm Bering Sea.

Neither of us was a good sailor and already Randall's skin looked pale and clammy, although it had been less than 10 minutes since we said good-bye to Milt and Chuck at the dock.

The wheelhouse was a quiet hum of activity as the skipper, Scott Bowlden, navigated through the narrow channel leaving Chernofski Harbor. He sat in a high pedestal chair on the right side of the room, with flashing screens and monitors all around him.

Over his head was the marine band radio, and Peggy Dyson's voice filled the small space with the evening marine weather. I listened to her give our Area 12 forecast and quailed. We were in for a northwest gale.

Scott turned off the radio and set his course, putting the boat on automatic pilot. "If she's right, it'll be a rough ride till we get around Cape Kovrizhka," he said.

I watched the calm water streaming past the boat's sides and prayed that for once she was wrong and we could make the 10-hour trip to Unalaska village without getting tossed all over an angry sea.

Right now it was beautiful. The placid water stretched as far as we could see, like a lake of mercury under gray, cloud-filled skies. Cormorants and gulls cut slashes through the air and settled like down on the shimmery surface. It seemed inconceivable that this benevolent, lake-like expanse could change in an instant to a boiling caldron of menace.

I crossed my fingers. Cape Kovrizhka was 35 nautical miles to the northeast, a monster headland rearing its way into the sea from the slopes of Makushin volcano. After listening to sailors for almost 3 years, I knew the waters around the cape were treacherous; full of riptides and undercurrents. I breathed a brief fervent prayer that the wind would hold off until we rounded the cape.

But it was in vain. Randall and I stood together, not speaking, both fighting increasing queasiness, while we watched the silver water wrinkle like aluminum foil into short, choppy waves as the wind began its banshee wail through the boat's antennas and masts.

Scott's 18-year-old son, Jay, bounded up the iron stairs from the galley. "Hey, you guys, want some ice crea...Whoa, kid." He stared at Randall. "You don't look too good. In fact, you look just like I did the first time I got on this boat."

He gave Randall a sympathetic grin. "You want to lie down? There's an extra bunk in my stateroom."

"Thanks." Randall swallowed and tried to smile. "I don't feel too bad, yet."

"You look like you're gonna puke any minute," Jay said bluntly. "You better get close to a bucket."

"Yeah," Randall sighed. "It's really rocking now."

Instead of turning away from the window, he stared at the horizon as he shuffled to the stairwell and lowered himself over the edge, all without turning his head. "You coming, Mom?"

CALM BEFORE STORM. *Taking a boat ride to Unalaska village would be all right if the weather stayed like this. But as Cora and Randall found out, heavy seas quite often make landlubbers wish they were back on solid ground.*

"Not unless you need me." My only hope lay in staying upright in the relative coolness of the wheelhouse. I climbed up into the empty pedestal chair on the opposite side of the wheelhouse from Scott so I could see the horizon better.

The rocking increased as darkness fell. I strained my eyes to keep in sight the faint line where sky met water. Soon even that was gone.

The *Valiant* was a big boat—to me anyway. It was over 100 feet long and quite wide. The wheelhouse was in the stern, so the foredeck, with its bright halogen lights, stretched out in front of us. It was all I could see.

The lights dipped and tilted at crazy angles as the boat rolled in an increasing trough. The northwest wind was hitting us on our port side, while the direction we were moving kept us wallowing between waves. When the boat rolled with the sea, the lights shined on the frothing water. I noticed with astonishment that the water was still dotted with seabirds calmly riding the crests of the waves.

I was constantly swallowing now to keep ahead of the water rising in the back of my throat. Determined to ignore it, I asked Scott, "How did you get involved in fishing in Alaska?"

I knew he was from Idaho, like myself. He was a big, jovial, gray-haired man, whose age could have been anywhere from 45 to 60. His seamed, weather-beaten face had the old-young appearance I'd grown used to seeing on Alaskan fishermen.

"Just bad judgment, I guess." His big, booming laugh rang out. "But I can't complain." He pointed through the wheelhouse window in front of him at the sea boiling up all around us. "She's been good to me."

"And now Jay is following in your footsteps?" I heard feet on the iron rungs behind me. "How long did it take him to get over being seasick?"

"This is the first trip I haven't been sick," Jay answered, his head and shoulders appearing above the stairwell. "I was ready to give up, too."

"He had a real good introduction to Alaskan fishing when he was 10," Scott chuckled, nodding toward the stateroom directly off the wheelhouse that he had offered to me earlier. "I brought him and his mother through Shelikof Strait, between Kodiak Island and the Alaskan Peninsula, during a storm much worse than this."

"Yeah, I can laugh now, too." Jay's ruddy, handsome face crinkled into a smile. "But at the time, it wasn't funny." He turned to me. "Mom tried to hold me in the bunk and Dad gave us a plastic bucket." He grimaced.

"Well, we fell out of the bunk on about the second wave and spent the rest of the time slamming into the walls with the bucket tipped over on us." He looked at me with a rueful grin. "That ended Mom's love affair with the sea."

I could understand that.

"How's Randall?" I asked him.

Jay shook his head. "Miserable. He wants you."

I knew if I moved, I would throw up. I just hoped I'd find the head before it happened. "Where is he?" I slid off the chair and weaved to the stairwell.

"Go past the galley," Jay called after me as I crawled backward down the iron rungs. "The last door at the end of the hall."

As soon as I reached the galley, I didn't need directions from the two crewmen laughing at a movie on the television monitor and eating huge dishes of ice cream—I could hear Randall from there.

A sudden lurch of the boat sent me sprawling against the wall. Wet nausea clogged my throat as I sidled along the hot, stuffy hallway.

I found him wedged into the bottom bunk, his head buried in a plastic bucket. Violent spasms shuddered through him. His whole body shook.

"Make me cold," he gasped between spasms. "Make me cold."

I lurched across the hall to the tiny head and wet a paper towel in the sink. It was stifling hot in the close room—too hot. Suddenly, I, too, got sick.

Eventually, I grabbed another soaking paper towel and crawled to his bunk.

His eyes were closed, his face ashen. I dropped the towel on-

189

to his face and went back for more, putting them behind his head and across his chest.

For hours I did this, until his hair was soaking and his shirt drenched. After I was sick, I had felt better. Poor Randall got no such respite.

Once, when the boat hit a slight lull and ceased its relentless pitching, he fell instantly asleep. But as the rolling began again, he opened his eyes with a moan.

"Promise me," he whispered between gasps. "Promise me we can take an airplane home."

"It's just the storm," I whispered back. "As soon as it quits, you'll be fine."

"It's never going to stop," he groaned.

"We're almost to the Cape," I reassured him. "When we go around it, the wind will be on our tail and we won't rock so much."

Just then, a huge shudder passed through the boat. We felt it roll onto its side. It stayed there forever before sluggishly righting itself.

My heart pounded. Could the boat roll completely over? I had heard of it happening before; of crew members trapped in their bunks, never even getting a chance to jump overboard. I looked at Randall's haggard face. Again the boat heeled over with such force I landed against the opposite wall. Randall's eyelids barely flickered.

Down the hall I heard the two crewmen laughing and then the refrigerator door slammed. Slowly the boat came upright and I slid across the floor to Randall's side. This must be normal or these guys wouldn't be laughing. If there was any danger, Scott would send Jay to warn us. I struggled across the hall for more wet towels.

By the time I got back, the boat's motion had changed. The roll was replaced by a short, dropping sensation, followed by a steep lift. I sagged against the bunk and wiped Randall's paste-colored forehead.

"We're around Kovrizhka," I whispered. "You'll feel better now."

His eyes fluttered open, his black pupils huge in the dim light, standing out in his white face like obsidian marbles.

STORM-TOSSED WAVES can whip up in a hurry on the Bering Sea, making boat travel uncomfortable, if not downright treacherous, for local fishermen. Randall and Cora leaned this firsthand on their stomach-churning voyage to Unalaska village.

"Only because I can't get worse," he croaked, reaching for the bucket that I had miraculously kept emptied enough that neither of us were wearing its contents.

On through the night the boat slogged, its wide stern riding up on the sea and settling into the trough when a wave passed beneath it, causing a sensation in my stomach like the downward plunge of a high swing ride; not pleasant, but tolerable. Even Randall's heaves lessened. After the first hour, he lapsed into an exhausted sleep.

I smoothed the wet, black hair off his brow, grateful that he was finally resting. His eyes had deep, bluish circles under them, his cheeks were sunken and his lips dry and cracking. He looked like...I had a sudden picture of him, bald and shrunken, clutching the metal rails of a hospital bed.

"No!"

I didn't realize that I had spoken aloud until he stirred restlessly under my hand. I patted his shoulder and he settled back down.

Every mother's nightmare jolted through my mind. A trivial sign; a routine visit to the doctor; a sudden mind-numbing statement and your whole life changes in an instant. I had seen it happen—seen parents with glazed eyes walking in a shell-shocked silence down dim hallways, putting their hands over their ears, closing their eyes, some even running.

But you could never run far enough or shut your eyes long enough. You always had to come back, sooner or later, to reality.

"No, dear God," I whispered. "Don't let it happen to us."

I curled into a ball on the floor with my heels braced against the opposite bunk and my arms crossed above my head and wedged against Randall's bunk. But I didn't sleep; always aware of the slow rising and falling of the steel hull beneath us, and the other menace, out there, waiting.

It wasn't until I felt the boat stop at Universal Seafood's dock in Dutch Harbor that I dozed off and didn't wake up until Randall shook me.

"It's 7:30, Mom," he whispered. "I'm starving."

I blinked my eyes open, and I felt stiff and disoriented. He flipped on the light and I squinted into his drawn features. His face was still haggard and white.

"How do you feel?" I mumbled.

"Empty," he said. "Let's go eat."

My stomach lurched uneasily as I rolled to my feet. "How can you even think about eating after last night?" I struggled into the head and slapped cold water on my face.

"As soon as we stopped rocking, I was fine," he chirped. "Hurry up. I want to get off the boat."

We found Jay in the galley. He told us Scott was still sleeping since he hadn't gotten to bed until 2, after tying up at the dock. We told him good-bye and asked him to thank his father. Then we slung the canvas duffels containing our sleeping bags and an extra change of clothes over our shoulders and scrambled off the boat.

Snow was piled everywhere, with only narrow paths to walk along. While Chernofski's gentle hills and valleys were already sprouting new grass, Dutch Harbor and Unalaska village were still in the throes of winter.

The difference was due to the steep mountains surrounding them. They crowded the small settlements so closely that I felt claustrophobic after our wide-open spaces at the ranch. And it felt like winter. Cold wind filled with icy sleet nipped our faces as we hurried inside the Unisea Inn, Dutch Harbor's only eating place.

The noisy bustle of people talking, feet stomping, chairs scraping, was disconcerting after our many months of quiet. We sat down at a small table facing the harbor and looked at the menu a harried waitress slapped in front of us.

"Better just have toast and milk," I advised as I scanned the selections, looking for juice. My eyes popped open—$5 for a glass of tomato juice. The same for orange.

I closed my menu and glanced around the room. Empty juice glasses littered the tables. One burly, bearded fisherman had two full glasses of tomato juice in front of him and was drinking a third. If crab fishing was on the way out, the news hadn't reached their pockets yet, I guessed.

"Are you ready to order?" The middle-aged waitress stopped at our table.

"Tea, please, with cream and sugar."

"Toast and a glass of milk," Randall added.

"That's all?" she asked over her glasses.

"We just got off a boat," I explained with a weak smile. "We're not very hungry."

"Rough trip," she sympathized, before hurrying away.

The tea was hot and sweet, but I was too queasy to enjoy it. Instead, I reveled in just sitting still, the floor solid under my feet, thinking about what to do now that we were here.

I knew only two women in the village well enough to call friends. One was Abigail Dickson, our fishermen friend Jim's wife, the other an Aleut named Kathy Grimnes. Both were handspinners and had visited the ranch.

Together they owned a small bookstore that also sold yarns, herbal teas and cards. They had taught me how to spin on a drop spindle made from a pencil and a potato, and I had developed a keen affection for them both. Before they left the ranch, they had issued a standing invitation to visit anytime we were in town.

Randall devoured his single slice of white toast and gulped a 4-ounce glass of milk in one swallow. "I'm still hungry," he said.

"We'll get some fruit at the store, then." I reached for the check. The toast and milk had cost $3.50. I shook my head. "Let's go."

Outside, we stood on the concrete sidewalk and looked at the swirling sleet. Dutch Harbor was on Amaknak, an island connected to Unalaska Island by a new bridge. The village clinic was across the bridge and on the outskirts of the native settlement—too far to walk in this weather.

Six rusting, mud-spattered vehicles, all with "Taxi" painted on them someplace, lined the sidewalk. We climbed into the closest one, an elderly gray van, and asked the long-haired, tobacco-chewing youth behind the wheel to take us to the clinic.

For all his appearance, he drove carefully and kept up a cheerful conversation, asking us what boat we worked on.

"We're from Chernofski," Randall said.

"Never heard of it," he replied. "I'm from New York, myself."

The van rattled over a rutted, gravel road, across the bridge, up a steep hill, then rolled down through the small village of Unalaska, a distance of roughly 2 miles. He stopped beside a low, gray painted building. "That'll be $10, ma'am."

I swallowed an astonished gasp and paid him, adding a $2 tip—probably an insult, but severely depleting my small cash reserve.

Lugging our bags, we opened the clinic door and stopped short. The small waiting room was jammed. All the chairs were taken. People sat cross-legged on the floor or leaned against the walls. Some were asleep, some had rough bandages on fingers, hands and faces.

Involuntarily, I covered my nose and mouth with a mittened hand. Who were all these people? They looked like disaster victims. Leaving Randall by the door, I picked my way through the mass of humanity, trying not to step on anyone.

I reached the window and waited until the receptionist finished her phone conversation.

"Is there a doctor here?" I asked the top of her bent head.

"Uh-huh," the woman looked up. "An osteopath."

Relief rushed through me. "Can I make an appointment?"

She glanced behind me at the throng of waiting people. "Come back at 1 o'clock," she suggested.

Unable to resist, I jerked my head to indicate the crowded room. "What happened?" I asked. "Was there an accident?"

She laughed and shook her head, reaching for the ringing phone. "Just our usual crowd of cannery workers during the king crab season." She lifted her head, "See you at 1."

Outside, the sleet bit into us again. We had three and a half hours before 1 o'clock.

"Can I go to the store?" Randall asked.

I hunched my shoulders. "Let's leave our bags at Jim's first."

"Okay." Randall grabbed my duffel and started up the narrow, muddy road leading to the valley.

Jim and Abigail lived outside the village at the base of Nirvana Hill, one of the few level places in the entire area. Their house was in part of an old military barracks that used to be a latrine. Abigail had decorated the single room in cheerful Bohemian style, with rough plank bookshelves, covered cushions and a tiny sleeping loft, reached by a pole ladder.

When we got there, Jim was leaving. "Hey!" He gave me a quick hug, his long arms wrapping around me, his short reddish beard tickling my cold cheek. A Scotsman whose father had been a professor at the university in Fairbanks, Jim was a tall, rangy man with blue eyes and an engaging grin.

"What are you guys doing in the big city?" He clapped Randall on the shoulder.

I told him we needed to see the doctor and asked where Abi was.

"She's in Anchorage this week." He opened the door and waved us inside. "How are Milt and Chuck?"

"Fine. Anxious to get started on the spring work." We put our hands out to the small oil stove's faint heat. "It's sure a lot more spring-like down there."

"Yeah," Jim laughed. "The Shaler Mountains do that. We get snow and you get rain." He stowed our bags under the table.

"But we get sunshine and you get fog in the summer, so it all evens out." He rubbed his hands. "I've got to get down to the boat. Can I take you somewhere?"

"Carl's," Randall said, naming the village general store. "Milt gave me $20."

My head whipped up. "After what he said about a dead horse?"

"Oh, this isn't a loan," Randall said quickly. "He said that no kid who hadn't been to town for over a year should go with empty pockets."

"If I had known that, you could have paid your share of the taxi," I teased, warmed by Milt's generosity, but wondering if he, too, was more worried about the outcome of the trip than he let on. "We better not let it burn a hole in your pocket."

Jim dropped us in front of Carl's, where we were immediately surrounded by a pack of barking, snarling dogs, all huge, shaggy creatures the size of small horses.

"They're friendly," Jim said. He waved and drove away as we battled our way to the door.

The store was a mixture of old-time cracker barrel informality and blacksmith shop hardware. One glance at their limp, bruised produce convinced me we had to have that greenhouse.

I left Randall in front of the ice cream chest and went to the hardware section with Milt's list, hating my mechanical ignorance. After a half hour of searching, I gave the list to the clerk and let him find the items. I had no idea what they even looked like.

Armed with our packages, we braved the sleet and barrage of dogs once more. Following the beach road that ran in front of Carl's, we passed the Russian Orthodox church on our way to the bookstore. It had once been a beautiful building, but now looked dilapidated and shabby, with peeling white paint and tarnished green onion domes.

Since Russian ownership of Alaska during the 1700s, most Aleuts had embraced the Russian Orthodox faith and were devout churchgoers. But their small numbers did not guarantee the kind of income needed to keep the ancient building repaired.

When we entered the bookstore, its familiar perfume of incense and fresh roasted coffee filled our nostrils. Kathy Grimnes sat behind the counter. She was dark-skinned and attractive, with curly, black hair. She called a surprised greeting.

The youngest daughter of a church deacon, Kathy had been a child when World War II erupted and spent the war years at a relocation camp in southeastern Alaska, an abandoned cannery from which many of the villagers did not return. She had then moved to Seattle, married, bore three children, entered college when the youngest was in kindergarten, then returned to the village to give her children a chance to discover their own heritage. I admired her a great deal.

We visited and drank strong, sugar-filled coffee until time for our appointment, while Randall read the cartoon books Abi always stocked.

When we returned to the clinic, the only change was in the faces draped over the chairs and slumped against the walls. I signed in at the window and we took our place at the end of the long, long line.

It was 4:30 before Randall's name was called. The nurse wore blue jeans under a white lab coat. Her name was Edna and she treated us with a friendly smile, something I don't know if I would have been capable of after the day she had just gone through.

She looked at Randall's chest and left us in a small examining room. Another half hour elapsed before the door opened again. Dr. Ott was youngish, pleasant and visibly astonished after his examination.

"It's nothing," he said with a perplexed look in my direction. "Did you say you were a nurse?"

"I did." A huge grin split my face. "But I'm a mother first."

BLUE GOOSE. The Grumman Goose was a welcome change from a rocking boat on the trip back to the ranch. The airplane is also a reliable way of getting supplies to the island, although it's expensive. The lamb is supervising the unloading.

Chapter Sixteen

THE MIGHTY ANGLER

Much to Randall's joy, we charted the amphibious Grumman Goose to fly home...but only after a futile search for any fishing boat that we might hitch a ride on.

After a day of exploring the village, Randall was ready to go home to his horse, his dog and his gun. Even a pair of llamas, shipped to the island on the annual ferry by a couple of Jim's friends who wanted pack animals, didn't hold his interest long.

Then a heavy storm front moved in from the southeast and made any kind of travel impossible. We were both prisoners in Jim's small living quarters, and every conversation between us turned into a bickering contest.

Abigail's spinning books, coupled with 2 months' accumulation of mail, kept me occupied. But by the time the storm lifted enough for me to visit the stores, I felt restless and dislocated. And I was disappointed in the shriveled potatoes and bruised apples that lined the store shelves. They looked worse than what we had at the ranch.

The narrow muddy streets, piled high on either side with dirty snow, made me long for Chernofski's soft grass-covered paths. And the steep towering mountains surrounding Unalaska's harbor filled me with claustrophobia.

I knew Milton and Chuck would be worried about our long absence. After unsuccessfully trying to reach the ranch from Pan Alaska's cannery, I called Nancy Maloney, co-owner of Air Pac, and asked her to get us home as soon as the weather permitted.

After 6 days of waiting, we rushed to the airport with glad hearts when Air Pac's Goose pilot, Tom Madsen, called. Our only detour was past the post office, where we picked up the latest mail, including a large package for Randall that made his eyes positively light up. There was no time to ask him what it was.

I didn't like flying, but Randall loved it. He climbed eagerly into the co-pilot's seat and strapped himself in. I don't know why I felt safer in a boat than in an airplane. Perhaps because in a boat there was only one big worry—sinking. In the plane I had to worry about falling out of the sky, hitting the water, then sinking. Three-to-one...no wonder I liked boats better!

To make the ordeal easier, I hedged my bets as much as I could by only flying with pilots I knew and trusted. Tom Madsen was one of them. I looked across the aisle at my additional safeguard as the twin propellers spun into life.

Jim was also along. Solid, reassuring and unafraid, he was capable of dragging us from a sinking plane and starting a fire with two rocks when he reached shore. I had begged him to come with us.

I gave Jim what passed for a smile and pulled my coat over my head as the wheels left the gravel runway and the plane banked sharply to miss Mount Ballyho, one of those towering peaks.

We floated into the clear, still sky. A "window in the weather" we called it, a lull between storms, similar to the eye of a hurricane, when the wind stops for a few hours and everyone works furiously to repair damage from the storm just past and lay in provisions for the one coming.

As the twin engines throbbed westward, I ventured a peek out my window. Hundreds of feet below and off to our left loomed the Shaler Mountains. Under their heavy mantle of snow, the jagged peaks looked like miniature Alps. If Chuck had been with us, he would have repeated his oft-spoken dream, "Someday I'm going to find a way through those."

Looking down on the stark pinnacles, I hoped it was a dream he would outgrow.

We flew on down the coast where each jutting headland looked greener than the last. It was as if we'd left winter behind in Unalaska village and traveled westward into spring. The closer we got, the more excited I felt.

The bays passing below our wingtips stretched inland like

placid mirrors—Pumicestone, Kashega, Kuliliak, Aspid and, finally, Chernofski, gleaming like a twisted cobalt pretzel, melted and draped in lace-edged tongues between the black lava reefs.

The Goose skimmed the surface and nestled its rounded belly into the still water with hardly a splash. Propellers spinning, the plane turned and plowed toward the beach where two small figures, three leaping barking dogs and one small lamb waited. With a final roar, the squat plane burst out of the water and turned a tight half-circle on the beach, pointing its nose and propellers out over the bay in readiness for take-off.

BIRD'S-EYE VIEW. *The ranch as seen from the air seems a small and insignificant place. But once Cora and Randall landed, the feeling of warmth, safety and solitude filled them. Those things make their ranch significant indeed.*

Tom shut down the engines while Jim opened the hinged half-door and fitted the aluminum ladder into place. The minute I smelled the familiar scents of seaweed, barnyard and tundra that were so distinctly Chernofski, I had to blink back tears and swallow the sudden lump in my throat.

I fiddled with the seat belt while the others climbed out, giving myself an extra minute. Chuck knocked on the window beside my head. "Hi, Mom," he shouted. I watched through the spray-streaked glass as he and Randall scrambled up the beach with Randall's box between them.

"Here, let me help you." Milt's callused, work-roughened hands covered mine. I looked up into his welcoming green eyes and swallowed again.

"I'm glad you're home," he whispered.

"So am I," I murmured back.

He unbuckled my belt and brushed the back of his hand across my cheek before catching me against him in a quick hug. "Let's go."

Outside the place, we invited Jim and Tom up to the house for coffee. But Tom had a backlog of flights waiting and couldn't stay. So we thanked them and said good-bye.

They roared across the bay in a curtain of white water, lifted into the air above Chernofski Point and almost immediately became indistinguishable from the gulls and eagles flying in lazy circles above the water.

As soon as the plane was out of sight, Unalaska village and the outside world ceased to exist. My eyes feasted on the brown, sun-dappled hills at the perimeter of our private empire—Foggy Butte, Mount Aspid, the "gunsight", Observatory Point. I then shortened my view to take in the sturdy, weathered ranch buildings and lichen-encrusted corrals.

In this small pocket of rough, untouched beauty, time seemed suspended; almost to stand still. Peep-Sheep nuzzled his nose into my hand and the dogs crowded close around my legs. I breathed in a deep lungful of tangy salt air and whispered a brief prayer, "Don't ever let this change."

The spell was broken by the boys clattering over the beach rocks to help carry mail and the few supplies I had purchased.

PLACE TO GROW. Milt and Randall in the new greenhouse built for Cora using material from the old military buildings.

Back in the house, Milt had a Dutch oven of lamb stew simmering on the back of the stove. The kitchen gleamed; even the windows were washed.

"I helped!" Chuck announced.

"It looks *wonderful*," I praised.

"Wait till you see our big surprise." Chuck's eyes gleamed. "I promise you'll love it."

All through lunch I tried to guess, but neither of them would say. As soon as the dishes were done, Chuck led us behind the house.

Attached to the south side of the old laundry shed was a small glassed-in lean-to. "A greenhouse," I whispered. "You built me a greenhouse." Delight in my astonishment spread over their faces. "We've only been gone a week!"

"If you had been gone one more day, we would have gotten it painted," Chuck boasted.

I was speechless. Milt took me inside and showed me how the windows worked to provide cross drafts and how the raised beds could be half filled with gravel for drainage. He even had running water and electricity piped in via a green rubber hose from the outside kitchen faucet and a long extension cord.

"Now all you need are seeds," Milt said.

"I tried to get some in the village, but when I asked at the store, the clerk looked at me like I had lost my mind." I held my hand out to the sun coming in through the glass, surprised at the amount of warmth I felt, then added, "Jim said he would get his wife to pick some up in Anchorage, but I don't suppose we'll get them in time to plant this spring."

"At least it'll be ready when they come," Milt consoled. "And you ordered some from the catalog too, didn't you?" He cranked the windows shut. "The red salmon start running in June. We'll get a few boats in the harbor then. Someone is bound to bring mail."

Chuck stuck his head around the door. "Come on, you guys. Randall wants to show you something."

We closed the greenhouse door and followed him to the big warehouse in front of the finger dock. Randall was already there, his beaming face flushed with exertion and excitement. In front of him, on the floor, a fat inflatable boat emerged from a crumpled pile of red and blue vinyl as he pumped on a small bellows with his foot.

"My boat came," he gasped. "Isn't she a beauty?"

"Boat? What boat?" I eyed the unfolding craft with consternation. "When did I say you could buy a boat?"

"Last summer, remember? I showed you the picture." The last fold straightened out and the boat filled the center of the room. Randall pulled out the bellows tube and swiped at the hair hanging in his eyes. "You said okay."

I remembered vaguely that we had discussed him getting a rubber dinghy for floating down the creek. I felt the sturdy fabric. This wasn't a dinghy. It was a boat!

"I thought you said it was only 6 feet long," I protested. "This is at least 10."

"Six feet is the inside dimensions," he rushed to explain. "It just looks big because the sides are so fat." He turned to Milt for help. "That makes it safer, doesn't it?"

Milt inspected the boat, squeezing the firm air-filled pockets and checking the air nozzle. "Looks safe to me," he said.

"Hey, Randall," Chuck hollered from across the warehouse where he was poking in the discarded box. "They even sent paddles." He brandished two aluminum oars in the air. "Let's go try it out."

"Not so fast." I grabbed Randall by the shoulder, feeling his taut resistance all the way up my arm. "Get on life jackets."

"Aw, Mom, do we have to?"

"Here," Chuck threw Randall a foam vest and shrugged into another. "Put it on before she decides to tie ropes around us and hang onto them from the beach, like she did when we first came up here."

I heard Milt smother a laugh as he helped them get the boat out of the warehouse door and down to the water. I marched out onto the dock and took up my lifeguard position, undaunted by their ridicule.

The Bering Sea didn't hand out second chances. The natives even had a word to explain its harsh lessons that meant "the quick and the dead". There was no room for carelessness. Either you respected the water or it killed you.

"Any funny business," I warned from my perch, "and out come the ropes."

They hopped in and Milt gave them a shove. Both paddled like madmen to put as much distance between them and me in the shortest possible time; afraid I might change my mind and tether them to the shore. Randall's clear voice carried across the water, "Good-bye, Attila!"

Milt joined me on the dock and we watched the erratic progress as they tried to coordinate their rowing.

"Can they tip it over?" I asked.

"It would be hard," he said, just as Randall leaned far over the side, making the boat lurch crazily. "But anything is possible," he amended, with a shake of his head.

"I saw a fish!" Randall shouted. "Let's get a pole."

They splashed their oars in the water and the boat jerked and weaved to shore. Chuck jumped out.

"You go fishing by yourself," he said as he tied the boat to a dock piling. "You almost dumped us out. No telling what you might do if you actually catch a fish." He jerked off his life vest in disgust.

"Milt, should we let him fish out of that thing?" I asked uneasily.

Milt shrugged. "It's flat calm. We're right here. He has to learn sometime."

Randall flashed me a triumphant grin and raced up the bank for his pole. When he returned, Milt walked to the boat with him. "Okay. No more leaning over the side. Keep your weight in the middle of the boat and don't stand up." He untied the boat and held the pole while Randall stepped in. "Stay close to shore, and if the wind comes up at all, you get back here."

Randall nodded. "I promise."

"Good." Milt pushed him off and we watched him paddle away. He knelt in the midsection and used both oars. After a few awkward motions and sloppy splashes, he got the feel of the oars and skimmed the boat across the water.

"That's far enough," I shouted when he was 50 feet offshore. Even that close, the boat looked tiny against the water; a small red dot in an immense stretch of blue.

Randall pulled in the oars and dropped his line overboard. After a few minutes, Chuck said, "This is boring", and disappeared inside the warehouse. I sat on the edge of the dock with my legs swinging over the water and pulled my jacket closer around me.

Another 5 minutes went by and the line hadn't moved. Milt straightened up from his leaning position against the winch and dusted the sleeve of his coat.

"I have some things to do, too," he said. "Holler if you need me. I'll be in the warehouse." He sauntered up the dock, leaving me to my solitary vigil.

Sometime later, Peep-Sheep came around the side of the building and clambered over the planks to nuzzle against my shoulder, telling me it was time for a bottle. Behind me I heard Milt and Chuck laughing through the open warehouse door.

I stood up. "Time to come in," I called to Randall's determined, hunched back.

"In a minute," he yelled across the water, reeling in his line. "I might have something."

"Hurry up." I raised my voice just enough to let him know I meant business. "Peep's hungry."

I doubted he had anything hooked. But pogies aren't fighters, so it was possible. I tapped my foot and waited. Just as I was ready to holler again, Randall screamed and fell over backward, disappearing into the bottom of his boat as it jolted out of the water and splashed back down with a geyser of spray.

"Milt, come quick!" I called.

Two sets of feet pounded down the dock behind me. "What happened?" Chuck's curious face appeared at my elbow, with Milt a step behind him. "Did he fall overboard?"

"No!" I shrieked. "He fell down in the bottom of the boat. Something hit it."

"Hit it?" Milt sounded skeptical. "Are you sure he didn't just lose his balance and fall down?"

"Yes, I'm sure," I insisted. "Look." Randall's head appeared above the fat balloon-like side of the boat. He clutched the pole with both hands. Across the water I heard the faint hiss of line unreeling at high speed.

"It's a halibut!" he shouted. "I saw it!"

"Hang on!" Chuck shouted.

"Lock your reel!" Milt thundered.

"Let go!" I screamed.

Randall crouched in the bow with the tip of the pole pulled under the water. I watch him reel furiously. Slowly, almost imperceptibly, his boat started to move.

"What's happening?" I whispered. "Why is he moving?"

"The fish is pulling him!" Chuck exclaimed, his voice high and excited. "Reel, Randall, reel!" he shouted. "Don't let him get away!"

"Throw that pole overboard, Randall!" I screeched, grasping Milt's arm. "Tell him to stop," I begged. "He can't hear me."

"Cut your line, Randall!" Milt shouted. "Come back to shore."

But by this time, Randall was a marble-sized dot halfway across the bay.

"Rats!" Chuck smacked his fist against a piling. "I should have gone with him. He isn't strong enough."

I watched in dumb fright as the tiny red dot made slow, lazy circles through the calm blue water; always away from shore, never toward us. Why doesn't he give up?

A cold, sinking feeling settled in the pit of my stomach. A halibut could weigh 500 pounds and be 10 feet long—far bigger than Randall's boat. And the minute their heads broke water, they went crazy. I had read stories about monster halibut breaking the shoulders of full-grown men; breaking their legs, too, and leaving them to die on the decks of their own boats.

"Milt," I whimpered. "Please go get him."

"Don't worry, Mama, he'll never tire out that big fish." Milt glanced at my petrified white face. "But I'll go get him, if you're worried."

"Just a minute," Chuck interrupted, his hands cupped around his eyes. "He's stopped." We all strained our eyes toward the red blot. "Yes. Now he's moving again. He's coming back. I can see the oars. He must have cut the line."

I tried to straighten out my clenched fingers, feeling a tight headache beginning in the knotted muscles of my neck. Hurry up, Randall. Get out of the water. Would the fish attack the boat? I had this sudden vision of red vinyl fragments exploding in a frenzy of white water. Hurry! Hurry! Hurry!

It must have been only minutes, but it seemed hours before I could make out Randall's face. It was set in a concentrated mask, as he pulled determinedly on the oars. As soon as he was close enough to hear, Chuck hollered, "What happened?"

"I forgot to tie the line to the reel," Randall groaned. "When he got to the end, it just came off."

Thank God! Thank God! Thank God! Inside I was one big joyful yell.

He paddled onto the beach and hopped out. Throwing the line around a piling in a careless half-hitch, he turned to us with a satisfied grin.

"That's the most fun I've had in my whole life!"

Chapter Seventeen

"WE CAN'T MAKE HIM ANY DEADER"

Mom, something's wrong with Peep." Chuck poked his head in the storehouse door where I was coating fresh eggs with Vaseline. Behind him I saw blowing snowflakes from yet another squall.

"He probably ate too much," I said, wiping my hands on my apron. "Between him and Tulip's new calf, the dandelion crop in the yard is just about decimated." I picked up a quart jar of cream I had been cooling and followed him outside.

Signs of spring greeted us everywhere. Raw, bitter wind touched our faces and a thin layer of ice pellets covered the ground. Yet, there was a difference.

Just above the western horizon, the sun hovered like a luminous orange basketball. I lifted my face to it while buttoning my sweater close around my neck. From the barn I heard the sound of Milt's post maul driving wedges into the pilings he was splitting into fence posts. It was 10 p.m. and no one thought about going to bed.

In the 6 weeks since Randall's clinic trip, our lives had speeded into a blur of preparation for the short summer of riding, shearing and gathering cattle that was almost upon us as May blustered by.

Peep-Sheep had grown into a fat, husky animal. He ate everything that sprouted through the ground, besides wanting his bottle of instant milk three times a day. Hearing our feet on the walk, he lifted his head and ran to the fence, the bell Milt had tied around his neck tinkling with every bounce.

"He looks fine to me," I scratched his ears through the woven wire. "What made you think he was sick?"

"Well...it's his..." Chuck gave me an embarrassed glance. "Uh ...his hangy-down part is all red."

209

MILT THE POST MAN. *Fence posts are made from treated dock pilings that wash ashore. The pilings are split, then sharpened, which Milt is doing here. It's a never-ending task, says Cora, as they never seem to have enough posts.*

"What?" I looked at him blankly. "Oh."

Chuck picked up the lamb so I could see his belly. "Good grief!" I gasped. "What has he gotten into?" His sheath was distended and bright red. "He was fine yesterday," I said.

"What's the matter with him?" Chuck asked.

I shook my head. "I don't know." Peep nibbled at my fingers until he got one into his mouth and sucked vigorously, butting his head against my arm. "He sure doesn't act sick." I pressed my hand into his abdomen; it felt soft and he didn't flinch. "I'll look in Milt's old sheep book. It might suggest a remedy we have on hand."

I turned to go back up the walk toward the house and Chuck opened the gate to the barnyard. "Where are you going?" I asked.

He shrugged. "I thought I'd walk out to the mouth and watch the waves, if Milt doesn't need me anymore."

They had been working since supper on the post pile, and before that, at my insistence, Chuck had studied all afternoon. "Go ahead," I told him. "But don't stay after dark, okay?"

He nodded and swung down the path, pushing through the

outer gate with a thrust of his shoulders against the pickets. I watched him go, sensing his restlessness. He walked the beaches and watched the pounding surf during all his free time lately. I sighed. What did I expect? He had just turned 17. He was growing up and he hadn't seen anyone his age for a year.

Could I bear to let him go? I curled my fingers into Peep-Sheep's wool with such intensity he struggled free. Chuck was my only natural child, born when I was 19. We had gone through a lot together. Would I ever be grown up enough to loosen those strings?

His slight frame disappeared into the barn while I thought about the faraway look in his eyes. Their preoccupation told me

ROCKY SOLITUDE. The reef is a source of endless adventure, with its starfish and sea anemones. And, as Chuck knows, there are always waves to watch. Val, Milt's youngest son who spent his entire childhood on Unalaska Island, is fishing from the reef.

I would have to let go, maybe not for a while yet, but soon. I picked up my jar of cream and took it into the house.

Randall looked up from the table where he was finishing his assignments. "Hey, good!" His eyes lit up at the cream. "Can I churn?"

Remembering how long it took me as a child to churn cream into butter, I had given Randall the task when Tulip came fresh, thinking it would keep him out of mischief for a couple hours at a time.

He took to the chore with his usual exuberance, though, and delivered a jar of butter in less than 10 minutes the first time he tried. I had been so surprised and lavish in my praise, he strove to shorten his time with every new jar.

"Sure," I said. "See if you can do it in 5 minutes."

I handed him the cream and went into our bedroom for Milt's sheep book. Copyrighted in 1950, it still covered most diseases and didn't rely on antibiotics so much for treatment. Usually, we could scrounge up the remedies it suggested.

Not really worried about Peep, thinking he had eaten something that was excreted in his urine, I was surprised to find four pages describing his exact symptoms, plus some very grim ones to expect in the immediate future.

Balanitis? I had never heard of it—an irritation which at this writing didn't have a known cause, but seemed to result from an accumulation of mineral salts in the urine that resulted in swelling, even blockage of the urine flow if the salts formed calculi. If this happened, Peep would strike his belly with a hind leg in response to the pain of a distended bladder. So far, he had not done that.

I read the pages again and again and zeroed in on the word "bacteria". If an infection was present—and it looked likely with all the redness and swelling—then sulfa might help him. As a nurse, I knew that some infections in humans were treated with sulfa drugs. Milt had a big bottle of sulfa pills the military had given him when they left the harbor. I wondered if it would work on sheep.

The porch door opened and I heard Milt talking to Randall. I took the book with me to the kitchen and got in on the last of their conversation.

"Six minutes, you say?" Milt raised an eyebrow at the jar of pale yellow butter Randall held up to the light.

"Yes," Randall boasted. "And I did it with one hand, too."

"Now, Randall," I chided. "Are you sure?"

But I was too worried about Peep to hassle him for exaggerating. I showed Milt the book and told him about Peep's symptoms.

"I've never heard of it," Milt said. "Where is he?"

The three of us went out in the yard and found Peep still eating dandelions like he didn't have a care in the world. Randall picked him up so Milt could see his underside.

"Holy cow!" Randall exclaimed. "What happened?"

"Looks bad," Milt whistled through his teeth. "Is he passing water?"

"I haven't seen him, but I wasn't paying much attention, either," I admitted. "Do you think sulfa would help?"

"Might," Milt agreed. "Shut him in the chicken feed house tonight so we can see if he goes. That's what worries me."

His eyes big with concern, Randall carried the protesting lamb to the building next to the chicken house and locked him inside. "Give him something to make him better, Mom," he pleaded.

"Don't worry, we'll do the best we can." I shivered against the raw night air, noticing as I turned for the house that the sun was completely down and Chuck hadn't returned.

Milt handed me his sweatshirt and I hugged it around me. "I'll get a bottle with sulfa in it." We walked back to the kitchen and they watched as I crushed half a sulfa tablet, calculating the dosage the same as if the lamb were a child, and put it in his milk powder.

"You really love that lamb, don't you?" Milt asked.

I nodded, unable to think about that frisky little woolly not getting well. "He's part of the family," I said.

"He isn't going to die, is he?" Randall raised startled eyes to mine. "I don't want Peep to die."

"None of us do." I handed him the bottle. "We'll give him this and hope he's better tomorrow." I turned to Milt and caught him in a tired yawn. "I'll turn off the generator when we finish," I offered. "Why don't you turn in?"

"Thanks. I'm beat." He squeezed my shoulder. "Don't stay up late."

In the chicken feed house, we scrutinized the floor in vain for wet spots. "I didn't think I'd ever be pleased to see you make another mess on the floor, Peep," I groaned. "But I'd sure be glad to see you make one now."

He gobbled the milk and sucked noisily on the empty nipple until Randall got it pulled out of his mouth. Then he nuzzled and butted against his hand, but he didn't jump and buck stiff-legged like he usually did. Was his walk getting a little clumsy, or was I imagining it? I shook my head and told myself not to make it worse than it was.

Sending Randall back inside, I walked slowly toward the generator building. I took my time switching off the current and listening to the deep chugging of the diesel engine die away. Afterward, I stood in the doorway and looked up at the stars, keeping Milt's sweatshirt hugged close against the cutting wind.

I had located the Big Dipper and the North Star before my ears picked up the sound they'd been waiting for—the gate latch clicking and Chuck's footsteps coming softly up the walk.

"What are you doing out here?" he asked.

"Looking at the stars," I said, pointing upward. "How were the waves? Any big ones?"

In the dark I couldn't see the shrug I knew was there. "A few." He opened the porch door for me. "How's Peep?"

"No change." I stumbled on the rug and he steadied me. Just that slight touch on my arm made my eyes fill. *I don't want you to grow up and go away.* For a second, I was afraid I had said the words out loud. "We're hoping he'll be better in the morning."

"Yeah. Well, g'night." Chuck glided toward his room, leaving me alone in the dark kitchen.

The next morning, Peep still hadn't passed any water, but the redness and swelling had almost disappeared. I was elated that the sulfa worked, but still, the lack of urine worried me. Was his belly getting bigger? He wasn't kicking at it yet. He drank his bottle eagerly and tried to dash past us into the chicken yard.

But Randall blocked him and locked the door. It was pouring rain outside so we wouldn't be able to see if he did go. I was just as glad to leave him inside. The book had suggested total fasting as a last resort, so I didn't give him any oatmeal as I usually did.

While the boys studied, I reread the sheep book, paying special attention to the section on probable causes. It sounded like diet was the most likely culprit. I decided to try Tulip's fresh milk again instead of using instant, even though the fresh milk had given him the scours the first time. Perhaps there was something in the instant powder that caused calculi. Perhaps not. At least it made me feel like I was doing something.

Between classes, one of us dashed through the rain and checked on him, never finding the puddle we hoped to see. When 4 o'clock rolled around and the kids put away their books, we took Peep a bottle of Tulip's milk with more sulfa in it.

The floor was dry. Peep ran up to us and drank the milk with his usual enthusiasm. When it was gone, he nuzzled my hand for oatmeal.

Then he raised a hind leg and kicked at his belly.

"He did it," Randall whispered. "He's going to die, isn't he?" He looked up at me, his eyes imploring. "You have to do something."

"What?" I asked helplessly, as I watched Peep bring his leg up in slow motion and kick at his belly again. "I can't make him go."

"What happens now?" Chuck asked.

"If he doesn't go, his bladder will rupture and he will die within 24 hours," I said, reciting words off the page I had memorized.

By the next morning, he still hadn't gone and he was kicking himself every few minutes. His abdomen felt tight to the touch and warm, despite the absence of inflammation. But when I offered the bottle, he gobbled it down and butted for more. I had cut him down to 6 ounces, hoping less fluid would at least make his discomfort easier to bear.

"I can't stand it," Randall moaned as he watched Peep stamp his feet and try to lie down, only to rise and continue his futile kicking. "Why doesn't Milt just put him out of his misery?" His bottom lip trembling, he bolted for the house.

It had come to that.

Alone, I squatted down beside the little animal and put my arms around his neck. Tears welled and splashed down my cheeks. "Oh, Peep," I cried. "If you were human, I'd put in a catheter and…"

As soon as the words were out of my mouth, I blinked stupidly as the impulsive speech sank in. Then I burst out laughing and kissed Peep on the nose.

"*Why not?*" I shouted as I stood up. "It sure can't make you any deader." I dashed through the rain to the barn where Milt was chopping kindling.

At first he was dubious. "I don't know." He rubbed his jaw. "I've never tried anything like that before."

"You can do it," I urged. "You've butchered lambs so you know right where the bladder is; all I want you to do is stick a needle in it and drain off the urine and relieve the pressure."

"What if I kill him?"

"He's as good as dead right now," I said. "I know it won't cure him, but he won't have to die in agony, either." I rummaged through the sheep supply cupboard for a 50cc syringe and an 18-gauge needle. "Come on, let's do it." Milt put down his hatchet and we splashed back to the chicken feed house.

I wanted to measure the amount, but the instant Milt penetrated the bladder wall, a geyser of bloody urine exploded into the syringe and shot the plunger across the small room. It was all Milt could do to keep the needle steady and in place. Even then it dislodged and fell out before the jetting stream had diminished.

"Shall I try again?" he asked.

I looked at the dripping walls and splattered floor. "There can't be much left. Does he feel softer?"

"Quite a bit." Milt ran his hand across Peep's belly. "He feels better, too."

The lamb twisted away from Milt and bucked stiff-legged across the room to his oatmeal dish. He didn't kick his abdomen once. "Let's wait till morning," I suggested. "If he hasn't gone on his own by then, maybe we can put in a supra-pubic catheter

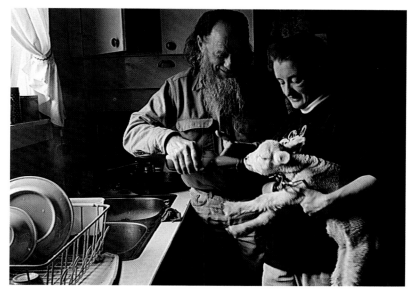

PET LAMBS were unheard of at this working ranch until Cora's arrival. Ever since she took in Peep years ago, some little lamb (like the one above) is part of the household. Milt apparently got used to the idea!

and IV tubing. It's in sterile packages."

Since I had spent so much time on Peep, I had to throw a quickie supper together. Meat was easy. I sent Chuck to the warehouse with a saw for chops off the hanging mutton carcass.

Canned pears, sprinkled with grated cheese and drizzled with sweetened mayonnaise, made a salad we all liked. Steamed rice substituted for potatoes, which we were now out of. Green beans sauteed for a minute in bacon fat tasted almost fresh, and hot applesauce topped with graham cracker crumbs took the place of a pie I had promised.

While I threw the meal together, I thought about how I could devise a workable catheter for Peep. Our successful bladder tap had given me fresh hope. It told me the obstruction was below the bladder instead of above it. A stone in the urethra might pass on out if we could keep him alive long enough.

The next morning Milt went with me to feed Peep. When we opened the door, Peep came running. It took a second to realize what I was seeing.

"I think his bladder ruptured," Milt said.

The lamb's belly hung to his knees. It swayed back and forth as he ran. It was then I gave up. We had done all we could. I was ready to ask Milt to put him down. Peep bleated and butted at the bottle, forgotten in my hand. In amazement, we watched him drain it.

"I can't believe it," I gasped. "He doesn't act sick. The book says they get dull and listless and lie around. I don't understand it."

Milt ran his hand under Peep's abdomen. "Well, he hasn't passed water, and it's obvious his belly is full of urine; it's only a matter of time."

"Shall we put him down, then?" I asked.

Milt scratched the lamb's ears and Peep rubbed against his hand. "I hate to," he sighed. "I guess as long as he isn't in pain, we can just wait and see what happens."

I let him outside and he wandered across the yard to Tulip's calf and began nibbling grass alongside her. At noon Randall and I took another bottle to him and the fluid in his abdomen had increased so much his bell collar was choking him.

He tried to bleat when he saw us, but all that came out was a strangled croak. Randall pried the buckle open and took off the collar. Immediately, Peep drank his bottle and followed us back to the door, his swollen belly swinging like a pendulum between his legs.

After supper, we fed him again and locked him in the chicken feed house. "Will he ever get better?" Randall whispered. "I don't like seeing him this way."

"I don't either." I noticed with despair the unnatural fatness of Peep's face. In the few hours since we had removed the collar, the fluid had seeped into his facial tissue. I was sure he would be dead by morning.

But when I opened the door after breakfast, he got slowly to his feet and lumbered across the floor to take his bottle. His eyes were slits in his moon-shaped face. Still, he drank eagerly, wagged his tail and followed me to the back of the house.

Chuck met me on the walk coming from the barn with a brimming pail of milk. "Milt says we should take the day off,"

PEEP AND PAL. Peep not only gets along with Milt and Cora and the boys, here he's photographed nuzzling Simon the cat. The horses, dogs and cats all have jobs on the ranch. Peep's job is just being cute.

he said. "The wind is down and the bay is calm."

Immediately suspicious, I asked, "Where is he?"

"Getting the boat in the water," Chuck grinned. "You better hurry."

COAL MINERS. The coal is there for the taking, left behind when the military abandoned Unalaska Island in the late '40s. Unfortunately, there's no delivery! Milt, Cora and the boys had to dig the coal out of this huge pile, sift it through a screen, bag it and haul it home by boat. Coal that requires this much work warms a body more than once!

Chapter Eighteen

ETERNAL VIGILANCE

Milt didn't take days off. For him to suggest doing so astonished me. I charged across the horse pasture toward the marine railway, wondering what he had in mind.

I didn't have to wonder long. As I turned the corner of the winch house, I bumped into Randall, who was struggling with two bundles of used feed sacks. Lifting one off his shoulder, I tightroped my way down the skinny track and tossed it into the boat, where Milt was stretched over the engine well, putting the plug in the bottom.

"Sacking coal is not a day off," I informed him.

"A change is the same as," he joked. "And we can beachcomb while we eat lunch."

"Oh, Milt," I made an exasperated face, "don't you ever play?"

"Summer's coming, my dear." He smiled up at me from his awkward position. "We have to take advantage of every decent day. There won't be many."

Randall trudged up the track and rolled his burden over the bow. "A day out of school is a day off to me," he gasped. "I like sacking coal. Besides, Milt promised I could drive the Bobcat." His face split in a huge grin.

"Yeah!" Chuck followed Randall up the track, lugging an extra can of outboard gasoline. "And Milt said I could drive the boat."

"I don't mind sacking coal, either," I protested, looking at the three expectant faces. "But don't you wonder sometimes what it would be like to take a whole day off and do nothing?"

They looked at me like I was crazy. "Never mind," I sighed. "I'll go pack a lunch."

While I was back at the house, I locked Peep inside the feed shed, not wanting him to wander off and die where I wouldn't find him before the eagles did. Then I made a thermos of tea and some peanut butter sandwiches, and stuck in the last of the oranges I got from the village.

When we got ready to shove off the marine railway, Milt sat beside me on the bow seat, leaving Chuck to start the engine and back away from the dolly. His face a mask of concentration, he accomplished, in less than a minute, all the start-up tasks.

From repeated pulls on the starter cord, to adjusting the fuel flow, to untying the rope that kept us fastened to the rails until the motor was going, he copied Milt's actions so perfectly, I knew he had them branded on his brain.

When he had successfully reversed away from the tracks, he shifted gears and moved the tiller to the left, bringing the heavy boat around in a wide, sweeping arc. Only then did I see him relax. He patted the tiller twice with his gloved hand and gave us a satisfied grin.

As we plowed through the still waters, gulls lifted off the surface and screamed above our heads. Squat black puffins with bright-orange beaks splashed and skidded clumsily between flocks of ducks in an attempt to evade our noisy monster of a boat.

Chuck deftly maneuvered between floating ropes of bull kelp and the lacy fronds of ribbon seaweed that grew from the sea floor 35 feet below us. I looked over the side into the dark, opaque stillness and wondered at all the life it hid, sheltered and supported, so close to our doorway, yet so secret and unknown. Anything could be down there in that vast, watery universe.

A sleek, round head broke the surface and the dark, unblinking eyes of a harbor seal followed our progress. As we rounded the lava rocks on Buck Headland, Milt pointed out two gray-headed sea otters that sat up and scolded us in high, strident cries, before diving under with a fluid twist of their rounded heads.

"There," Milt said as we tied the boat under the slaughterhouse dock. "Wasn't that a nice break?"

"Yes," I agreed. "Do you care if I walk to the coal pile instead of going in the boat?"

"I feel like walking, too," he said. "Chuck, take the boat across Mutton Cove to the dolphins, okay?"

"No sweat." Chuck sat back down and, with barely concealed

impatience, waited for Randall to jump out. Before we got up the steep beach, he had the boat reversed from under the dock and purring across the cove.

"He does that very well," I commented, watching his smooth wake.

"I usually let him drive when we come across," Milt said in an offhand tone. "He likes to, and the practice is good for him."

"Hurry up, you guys," Randall called. "Open the door. Chuck's gonna beat me."

Inside the slaughterhouse, while Milt watched, Randall filled the Bobcat tank with diesel fuel, checked the oil and hydraulic fluid level, then started the engine.

After a quiet reminder not to hot-rod, he chugged off down the only road on our end of the island, a 2-mile dirt lane connecting the two big military docks on opposite sides of Mutton Cove.

I watched him disappear around the first corner. "I can't believe the change this past year has made in my kids," I said.

"Our kids," Milt replied, reaching for my hand.

"Thanks," I whispered. "Not many men would have taken on that responsibility."

"I'm the thankful one." He squeezed my hand and we started down the rutted road. "You three filled a very big void." He met my eyes as he said the words, unafraid to let me see how much he cared. "And I'm sorry we can't take a whole day off. I hate being practical, but I have to be."

"This is a perfect day off," I smiled. And I meant it. We splashed through the water-filled ruts in contented harmony, hands swinging in a loose clasp, our boots sloshing in unison.

The road had been cut out of hills that sloped to the shoreline. Light gray stone gleamed wetly along their exposed banks, and out of every crack, wildflowers grew. Great clusters of white anemones sprang from narrow crevices, while huge circular mounds of yellow cinquefoil clung to the rock face itself.

Small clumps of fern and spreading lupine pushed their greenery among the brilliant blooms. I stumbled along the rutted, muddy road, letting my eyes drink in the soul-nourishing vista, knowing that if I lived 1,000 years, I would never tire of this humbling display of earth's renewal.

When we reached the coal pile, the boys had unloaded the sacks and Chuck had one fitted over the chute end of the slanted, 10-foot coal screen. We all took our places in the assembly line process Milt had devised to gather the best coal in the least time.

Randall clawed his way to the top of the screen with a shovel, while Chuck stationed himself below Randall at the chute, ready to work the trap door that released coal into the sack. Milt climbed into the Bobcat, lowered the bucket and scooped out a bite of coal. I unraveled lengths of rope salvaged off the beach for ties while I waited for the first sack.

What we called the coal pile was really a 20-foot face, reaching 50 feet along the base of Ram Point, a cape running into the Bering Sea. Before the military left the harbor, they had bulldozed several feet of rocks and sod over the huge mound. Now, 50 years later, we were still sifting out rocks the size of cantaloupes.

With the Bobcat bucket, Milt scooped up coal and dumped it onto the screen. Randall scattered it across the mesh and Chuck worked the trap door, both hands pitching out rocks that plummeted through the opening along with the coal. Chuck tried to guess as close to 50 pounds per sack as he could. Even that made a heavy burden to lift, load and transfer for the four more times we would have to handle each bag before we had them across the harbor and into the coal shed.

As soon as he jerked the bulging bag off the chute, I grabbed it and pulled it out of the way while he fitted on another sack. While it filled, I tied the full sack shut with a piece of rope. When we had 10 sacks done, Milt and Chuck loaded them into the Bobcat bucket, drove them to the beach and dumped them on the bank.

After the first half hour, we were all sweating and thankful for the crisp, cool air. It was hard, dirty work and we were soon covered with coal dust, thirsty and hungry. When Milt loaded the second 10 sacks into the Bobcat for the trip to the beach, he took off his gloves and wiped the perspiration streaming down his face.

RENT FREE. Coal wasn't the only thing the military left behind. Most of the buildings on the island are old Army structures that were abandoned in the late '40s. When the buildings are no longer usable, their boards are used for ranch projects.

"This coal generates a lot of heat before it ever gets in the stove," he chuckled. "Must be magic."

When we had 50 sacks on the beach, we ate our sandwiches, sitting on driftwood logs with our grimy faces to the sun. Afterward, the boys and Milt beachcombed along the sea side of the isthmus while they enjoyed their oranges, section by section.

Knowing Randall would be starving by mid-afternoon, I saved half my sandwich. Pouring myself another cup of tea, I dropped an orange rind into it and sipped the pungent brew while I basked in the warmth and rested every one of my sore muscles by trying not to move even an eyelash.

From where I sat beside the ruins of an enormous marine way that had extended out into Mutton Cove, I could see the outlines of the old military roads snaking over the hills. One curved around to the east behind Ram Point and went past a jumbled collection of boards that once housed a base theater. Another was graded along a hillside above the old village midden and

DRY WELL. This abandoned military hydrant came all the way from an Iowa foundry, according to its raised lettering. These days, there's no fire department at Chernofski.

led to the remains of gun emplacements out on Chernofski Point, where the rusted barbed wire still stretched just under the summit and caught an occasional sheep.

Along the beach road in the distance I saw rotted wooden steps zigzagging up the steep banks to Quonsets half buried in the tundra, some still wearing a rusty coat of armor even after 50 years.

I thought about the men who had lived in those metal shells; tried to visualize what it must have been like then, with 4,000 people here instead of four. I imagined dark-green trucks roaring past with mud jetting from their tires, splashing on young men with crew cuts and smooth faces that hardly needed a razor; who all looked like Chuck.

I could see the harbor full of ponderous ships and I could hear the noise. But I could not make real the fright I knew they must have felt because I had never experienced it; that actual fear for your life at the hands of another human being.

I looked at the haphazard mounds of silvery boards that had been barracks, at the single, incongruous fire hydrant that leaned patiently beside the road, at the rusted iron cannon barrel that still pointed toward the sky, and felt a warm surge of kinship with all those men who would forever be strangers to me and to whom I owed such a deep debt of gratitude. Whether they wanted to be here or not, the fact that they were meant that I didn't have to be afraid.

"You're looking awfully serious," Milt said over my shoulder. "What are you thinking?" He reached for my tea cup.

"About how much I love it here," I told him, pouring the cup full of tea again. "About how peaceful it is, and safe."

We shared the steaming cup until we saw the boys. Then

we walked back to the coal pile. As we passed one of the fallen buildings, I stopped and pulled out a weathered gray board.

"This would make perfect paneling for the bathroom," I said.

"Those rough old one-by-sixes?" Milt looked at me in surprise.

"Exactly," I said. "With your clawfoot tub and wall-hung sink, it would be pure frontier Victorian."

"Uh-huh," was his only answer.

The second 50 sacks seemed like 200 by the time we finished them. But the weather held and the tide came in, so we didn't have as far to lug each sack to the boat. All of us were in high spirits as we dragged and pushed the clumsy bags into the boat.

Chuck and Milt worked on their own. But Randall and I car-

WARM WOOD. Planks from a fallen building made perfect paneling for Cora's cozy "frontier Victorian" bathroom. Scavenged material from abandoned military installations makes living a bit easier in the Aleutians.

ried one sack between us. Still, we finished in good time and set off back across the bay with plenty of daylight remaining for the 2-mile trip home.

Since the bay was so calm, Milt allowed 40 sacks aboard. They took all the available space except for a tiny spot for him to steer. The boys perched on top of the mound and I sat on the bow. The load was so heavy that after we pushed off, the water came right up to the gunwales and I could look almost straight across at the smooth water streaming past our pointed nose. As we came around Observatory Point, the sun was low enough in the western sky to send a gleaming, golden swatch of light across our path.

I watched it ripple over the water and thought how beautiful it was, and how nice a day it had been. The air was so still and calm; warm without a hint of rain. I stared dreamily into the water, mesmerized by its silky surface, thinking again how wonderful it felt to really relax my eternal vigilance for once and just enjoy the boat ride without worrying about wind, waves, ice or logs. After all, what could possibly happen on a millpond like the bay was tonight?

With a sigh of pure bliss, I looked into the clear, silent depths

CALM CROSSING. Cora does not often operate the dory that is used to ferry coal across the bay, but it's a calm day and she's trying her hand. Cora freely admits she's a better cook than a sailor.

...and saw a killer whale the size of New Jersey dive under our boat! For a moment, I was shocked into immobility, thinking only in a detached way how much bigger it was than our puny boat, how thick and massive it was—like the Goodyear Blimp—and so fast! Almost before I whipped my head around, it had surfaced on the opposite side with an ominous sound of rushing water.

Seemingly without effort, the huge black and white shape lifted out of the water in a graceful ballet leap and splashed back with a resounding crash that rocked our boat and sent spray cascading over us.

It happened so fast I couldn't scream. I felt as if I had forgotten how to breathe. Behind me, the boys were silent. I felt the boat turn in sluggish obedience to a change in the rudder. Without fanfare, Milt guided the craft out of the channel and the whale's path.

Clinging to the bow, I watched the dark, triangular dorsal fin glide through the water. Suddenly, it was joined by two smaller fins. Then with an eruption of spray like an exploding bomb, all three black glistening bodies shot skyward, soared into the air in graceful arcs and slapped back into the water with a noise like rifle shots.

"Wow," Chuck whispered. "Wow."

"Wow," Randall echoed. "Wow."

No one else said anything. We watched in silence until the three fins were tiny specks out in the channel. Then Milt nosed the boat toward home once more.

Reaction set in and I began trembling. The whale had been so menacing and so beautiful at the same time, I couldn't decide whether to bewail the tremendous fright it had given me or delight in the incredible experience. By the time we reached the beach, I had done a bit of both.

Frightening and exciting beyond description, the incident was one I knew I would never forget. By the looks on the boys' faces as we landed, I doubted they would either.

But no matter how shaken or excited we were, it was almost dark and we all hurried to get chores done. While I dashed to the house and stirred up the fire, Milt and Chuck unloaded the boat and Randall fed the chickens and let Tulip in the yard to feed her calf.

Just as I was breaking eggs into a bowl for a cheese and bacon omelet, Randall came screaming through the screen door.

"Peep-Sheep is going!" he shouted. "I saw him!"

Chapter Nineteen

THE GOOD SAMARITAN

In June sunlight on the last day of school, I followed Peep's telltale path. For the past 2 weeks, he had dribbled a steady stream wherever he traveled.

The obstruction had destroyed his control, but the needle puncture had saved the bladder from rupture by allowing urine to leak into his abdominal cavity.

The pooled urine caused the wool to fall off his underside, leaving bare patches of thin new skin. But we were so happy to have him well, we wouldn't have cared if all his wool fell off. I found him squeezing his head through the fence Milt had built around our budding daffodils, the only stateside flower we successfully grew.

"Forget it, buster," I threatened. "I've waited a long time for those to bloom."

As Peep sucked on his bottle, I admired the nodding heads, both anticipating, and dreading, the opening of their brilliant yellow blossoms. Daffodils signaled the beginning of summer, and that meant gathering sheep on horseback.

I sighed. It wasn't the sore muscles, the bone-jarring punishment of 18-hour rides or even my uneasy truce with the horses that depressed me. It was my ineffectiveness.

I hated it; being so useless, always behind, never in the right place. To be brutally honest, I was a liability on rides. Someone always had to watch out for me.

Well, I could change! My lips formed a straight line as I pulled away Peep's empty bottle. This summer I was determined to carry my own weight, be a real help.

"I'll do it," I vowed. "Even if it kills me."

During the afternoon, as I helped the boys put finishing touches on their lessons, I dreamed about galloping fearlessly across the tundra. By 3, Randall sealed his bulging envelope with a war whoop.

"Yahoo! No more school! I'm done! I'm done!" He dashed away, making a beeline for the barn, where Milt was sharpening shearing combs and cutters.

Chuck was quieter. He stuffed his last two exams into envelopes and scribbled his return address in the corner.

"Well, that does it," he announced. "I'm out of school."

"For the summer, anyway," I qualified, his tone of voice driving all-equestrian-of-the-year fantasies from my mind.

"No, Mom, this makes 12 years."

"But you're not finished."

He pressed his lips together in a stubborn line. "I'm quitting." He didn't bluster or shout. I could have handled that better than this deadly calm—a mature, adult calm. "I won't take more correspondence courses."

"**Why not think about it** over the summer?" I suggested. "You don't have to decide right now."

"I didn't decide right now. I thought about it all winter." He balanced his notebook on his head like a mortarboard and strutted around the table. "There, I graduated. Okay?"

So this was what had been going on behind his faraway expression; all those solitary walks and wave watching.

"You should graduate, Chuck." I made my voice sound reasonable. "You need that diploma."

He grinned. "Not to pound staples, I don't." With a flick of the wrist, he sailed the notebook into the coal bucket. "Milt wants me to check the lake pasture fence between Granite Creek and the head of the horse pasture." He looked at me with a mixture of defiance and pleading.

"Then you better get going," I said.

"That's all?"

"What do you want me to say?" I asked.

He shrugged, stuck his envelopes in the mail basket and walked out of the kitchen. A few minutes later I saw him leave the barn with his gun over one shoulder, a roll of smooth wire over the other and a hammer dangling from his hand.

Soon after, Randall came up the walk from the warehouse carrying a rusty stovepipe and dragging a ragged roll of wire.

He dropped them in the front of the old bunkhouse privy and came to the house.

"Mom!" He slammed the porch door and kicked off his boots. "Milt said I could use Val's old smoker, so I'm fixing it up." His face was flushed with excitement; summer vacation written all over it. He rummaged in our kitchen junk drawer until he found a pair of pliers. "Come look at it."

I followed him across the yard where two rusting 55-gallon fuel drums lay hidden in the putske bushes. He pulled away the broad leaves with a grunt of satisfaction. "All right. This is perfect."

One drum lay on its side with a crude door cut in one end, and the other drum was perched upright, 6 feet up the hill. It also had a jagged door cut in the top, but with air slits chiseled in the side and rough wire racks inside. The barrels were connected by disintegrating stovepipe.

"Isn't it neat?" Randall cried. He opened the lower door and it broke off in his hands. "See, you build a fire with cottonwood in here and you put your fish in the other one. Then the

RUST IN PEACE. There's a good view of the bay and the snow-dappled hills from these old tanks, just a couple of the items left behind by the military and put to good use by the Holmeses, as in the case of their improvised fish smoker.

HERDED SHEEP on winter pasture near Peacock Point will be driven toward fenced areas. Fence mending is a never-ending chore at Chernofski, as Chuck learned.

hot smoke goes up the stovepipe and fills the barrel and cures the fish." He gave me a gleeful smile. "Neat, huh?"

I agreed and he rubbed his hands together in a business-like way. "Well, I've got to get busy." He pulled the ruined stovepipe out of the way and began cleaning ashes from the barrel's interior.

Dismissed, I returned to the house. After putting four mutton shanks to boil with dried celery and bay leaves, I stoked the fire and went to the barn to tell Milt where I was going. Then I followed Chuck.

Granite Creek fence lay 2 miles southwest of ranch headquarters and marked the farthest fenced sheep boundary. The creek itself flowed west into Station Bay and served as a fence along the whole southern side of Peacock Pasture. The fence stretched southwesterly away from the creek and enclosed about 2 square miles of grass, lakes and high scree.

As I walked, I saw small bunches of sheep grazing in the deep grass. The bunches ranged from six to 35 animals, and whenever they caught sight of me, they took off at a dead run for the highest peak they could find, the lambs showing up as white flashes against the shaggy gray fleeces of the ewes.

I didn't know what I would say to Chuck, hoping something clever would materialize by the time I reached him. Instead of trying to marshal any kind of argument, I blanked it out and gave my whole attention to the rare pleasure of walking the gentle hills and valleys between headquarters and Granite Creek. Now that the temperatures were staying in the 50s and darkness didn't fall until after midnight, the hillsides were ablaze with color and the air full of bird song.

Lapland longspurs flew in short bursts in front of me. The bright chestnut flag behind the male's head singled him out from his less colorful mate. But both emitted a series of beautiful tinkling notes as they flew.

Among the rocks and scree on the crests, sandpipers scurried about, dragging one wing the moment I got too close. Their nests were so well hidden I had never found one. Even after I passed, their thin, shrill piping scolded me.

Lavender cranesbill covered the ground like a carpet. Blue lupine and yellow coastal paintbrush lifted graceful plumes of color out of its midst. As I climbed the hills where the grass thinned, pink and white fleabane reminded me of stateside zinnias with their wreath of pointed petals around a yellow clus-

ALEUTIAN WILDFLOWERS. Bright pink Kamchatka rhododendrons, left, blossom only for a short time, while the more subdued wild geraniums last longer. Japanese iris and Aleutian cotton, below, like to live along creek banks and other wet places.

ter. Iceland poppies nodded their lemon-colored blossoms among fat clumps of campion moss.

In the valleys, fireweed, the first flower to return after a forest fire, sent slender shoots through the grass and wild strawberry leaves. Edible and tender enough to earn the name "Alaskan asparagus", its green promise told me time was passing.

While daffodils announced summer's arrival, fireweed foretold its end. When the intense magenta blossoms appeared in August, we knew fall wasn't far behind.

I found Chuck at the top of the first hill coming away from Granite Creek. He looked up with a wry expression. "I've been expecting you."

"Oh, really?" I pretended innocence.

He gave the wire a final "twang" and started down into the gully.

"Don't forget your gun," I called.

"You bring it," he said over his shoulder. "It's a danged nuisance. I don't know why we have to carry one anyway."

"Because a wild bull isn't going to wait for you to run home and get it," I said mildly, sidestepping down the steep slope behind him with the gun held awkwardly in front of me.

He grunted as I leaned against a post. "As long as you're here, you might as well help." He tossed me a canvas bag of staples. "Hand me one when I ask for it."

"Sure." I shook some into my hand, wincing as the sharp double points dug into my palm. Chuck crouched on the opposite side of the five-string barbed wire fence so we faced each other as he worked. I held out a handful of staples and he took what he needed.

Since our winter had been mild, the fence damage was minimal. In the gullies where the snow had drifted, the breaks were the worst. Chuck fashioned loops in the broken ends, fastened a smooth wire splice to one and threaded it through the other, wrapping it around the hammer head and anchoring the loose end between the claws.

Then he turned the hammer, rolling the smooth wire around the head until the fence strand was taut again. Bending the wire back on itself, he secured it with a couple wraps and cut off the excess, enough for another splice. When he was satisfied, he re-drove all the loose staples, hunting in the grass around the fence post for any used ones he might find before taking one from me.

He noticed me watching and gave me a sideways grin that ended with a snicker. "I know. I know. Put the staple perpendicular to the wire." He pounded one in with exaggerated care. "And always leave room for the wire to move."

I knew what he meant. They never let me forget the first time I mended fence. I hadn't asked for instructions and Milt never dreamed anyone raised on a goat farm in rural Idaho could be ignorant of such basic knowledge. I had pounded staples right into the fence posts all around the little house pasture. The first heavy snowstorm

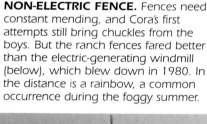

NON-ELECTRIC FENCE. Fences need constant mending, and Cora's first attempts still bring chuckles from the boys. But the ranch fences fared better than the electric-generating windmill (below), which blew down in 1980. In the distance is a rainbow, a common occurrence during the foggy summer.

had snapped wire right and left. The boys loved it.

"And you had a diploma," Chuck giggled.

Pouncing on the word, I jumped in. "What will you do, Chuck?"

"Huh?" He looked vague. "Oh, you mean when I grow up?"

"Yeah." Despite myself, I laughed. "Something like that."

"You know, it's funny." He took off his cap and wiped his face with his forearm, then gave me a puzzled look. "When I was in kindergarten, my teacher asked me that." He shrugged. "I didn't know what to tell her, either. The other kids said they wanted to be doctors and lawyers and firemen. But I didn't want to be any of those things. I still don't."

"You must want to do something."

"That's just it." He picked at lichen growing up the side of the post. "I don't want to be anything." He tapped his fingers against the top wire. "I just want to kick back."

"Kick back?" I said. "What's that supposed to mean?"

"I have no ambitions, Mom, and I don't care a lot about money." He pushed away from the post. "I want to explore Alaska, fish when I need a grubstake and help out down here when I'm broke."

"You're so bright," I faltered. "You could be anything you want."

"Good," he flashed me his most endearing smile. "I want to be a bum." He grabbed the gun. "Come on, it's late and we aren't half-finished. I don't want to get rained on."

We worked in silence for the next mile; me passing staples and him pounding, splicing broken wires and kinking stretched ones until they fit tight against the posts.

A bum! My son wanted to be a bum!

"Hey!" Chuck's sudden exclamation startled me out of my reverie. "What the heck?" He dropped the hammer and slithered down the steep slope on the outside of the fence, disappearing from my view.

"What is it?" I called.

When he didn't answer, I crawled under the fence and picked up the gun. The hillside was very steep. I sat down and was still almost standing up. My feet slipped on the slick grass. Cradling the gun in my arms, I slid to where Chuck knelt beside a cow that was on her side in the grass.

"I brought the gun, Chuck. Do you need it?"

"Not yet. Put it down."

He stood up and walked around the quiet animal. She looked young and thin, a long yearling or coming 2-year-old heifer. She wasn't having a calf and didn't look like she was about to. We stood staring down at her and she didn't move. Her one visible eye watched us without apparent interest. When I stooped and patted her between the horns, she didn't even blink.

"I wonder what's wrong?" I murmured.

"She must have slipped on this slick grass and fell with her head downhill," Chuck said. If we can get her straightened around, maybe she can get up."

"She looks heavy." I looked at him dubiously. "Should we get Milt?"

"Let's try first." He shrugged off his jacket and grabbed the tail. "Help me pull."

I gingerly grabbed her tail behind Chuck's hands.

"Okay," he directed. "Pull."

She was a lot of dead weight. But we were pulling her hindquarters downhill, so that helped.

"Again," Chuck groaned. "Puuull."

I felt her move. "Just a little bit more," I gasped.

It seemed like we struggled for hours. By the time Chuck was satisfied, my arms felt 18 feet long and sweat dripped from my eyes.

"Phew!" Chuck dropped beside me, wiping his hands on the grass. "Well, she's headed uphill again. She should be able to get up."

He rested for a few minutes, regaining his breath. The cow didn't move.

Chuck nudged her hip with his foot. "Last summer Milt and I got one out of a creek in Cattle Bay. I just can't remember what all he did." He walked around the still animal.

"Maybe if I get her on her brisket, she can get her legs under her." He knelt and pushed against her shoulder. Suddenly her head came up and she let out a strangled bawl. "It's working!" Chuck shouted. "Twist her tail!"

I grabbed and twisted. With a monumental surge, the cow lurched over onto her legs and staggered to her feet. She stood there swaying and blinking, with Chuck steadying her head and me holding her tail.

"Good girl," Chuck praised. "Good girl."

"Do you think she can walk?" I asked.

"Yeah, I think s..."

His reply was cut short, as a long shudder ran through the cow. She pulled her tail out of my hand and slapped me across the neck with it. I jumped back, but the second slash caught me like a whip around my shoulders. I scrambled backward up the slope.

All at once she whipped her head around and stared at Chuck, who was less than a foot from her. Her eyes rolled wildly, then she let forth an outraged bellow and charged.

"Run!" he shouted. "Get under the fence!"

He ran too, straight for the gun, with the cow snorting and bawling right on his heels. Shaking, I watched his lean body bend double and scoop up the weapon. Like a ballet dancer, he pivoted on one foot and raised the gun. He pointed at her and squeezed the trigger.

Nothing happened! She kept coming!

"Shoot!" I screamed. "Shoot!"

In slow motion I watched Chuck's arms go up in a desperate attempt to evade the murderous horns. I saw jets of mucous spurt from the cow's nostrils, heard her hoarse bellows of rage. Then, just as those horns touched his shirt, he sidestepped in a graceful matador feint. The cow galloped past and kept on going.

Chuck collapsed in a heap.

My own body folded up like a noodle and I slid the few steps to his side. His hands shook, his cheeks were pale and a white line ringed his mouth. I wiped the slime off his sweatshirt with my sleeve.

"Why didn't you shoot?" I didn't recognize my own voice, it trembled so.

"I tried," he croaked. The gun was still clenched in his

hand. "But I missed the trigger and hit the safety instead."

He flopped back in the grass; his breath came in deep rasping gasps. "Never again," he snarled. "I will never help another cow as long as I live. I don't care if they all die." He rubbed his hands along his arms. They were covered with goose bumps. "Ungrateful old bat."

For a few minutes we rested. We could see the cow far in the distance, still running, still bawling. After we got our breath back, we climbed the hill and crawled under the fence.

"You can sure run fast for a bum," I said lightly.

"Yeah, well, I'm not doing it again. That was my last Good Samaritan deed." He picked up the hammer. "Let's go home, I'm starved."

We hadn't gone 10 steps when raindrops splattered us. "It figures," Chuck groaned.

We hurried along the fence line toward the horse pasture. As we neared the west fork of Bishop Creek, I started downhill. I saw the second cow before he did. She was wedged on her side in the fast-flowing stream. On the opposite bank, a calf was curled in a tight ball. "Oh, no!" I cried.

When Chuck saw her, he threw his jacket on the ground and stomped on it. "No!" he shouted. "Not again. I refuse."

I didn't blame him. Both of us had just been scared silly.

Seeing us, the tiny calf got up and stumbled around on weak legs, shivering in the rain. Her plaintive cry floated on the breeze.

With a resigned sigh, Chuck plunged into the creek. "You wring on her tail, Mom. I'll get in front and make her mad."

WATCHING THE WOOLLIES. In spite of this serene beauty, there is danger for the unwary, as Randall found out while rounding up sheep. West Point, below, has steep cliffs.

Chapter Twenty

"W H O   A M   I ?"

A m I an Indian, Mom?" Randall reined in George beside Stormy as we followed the back of Milt's head along the now-familiar and much-used trail to Lamb Camp. This was the third time in a week we had gone after the same few renegades, and they got wilder each time we chased them.

I looked into his dancing black eyes and wondered what was behind that innocent face. In the 2 weeks since school vacation started, he had driven us all crazy with his constant chatter and practical jokes.

"I suppose it's possible," I said warily. "Why?"

"I'm kinda dark...'n'..." He hesitated, pulling back on George's reins. "And I'm different from you and Chuck. I don't even think like you guys do...so I thought I was an Indian... maybe an Apache?" he ended hopefully.

"Is that why you're wearing cowboy boots?" I teased.

"I'll tell you what I think." Chuck trotted up beside us and glared across me at Randall. "After yesterday, I think you're a jerk."

I stifled a laugh as the scene in the shearing barn replayed behind my eyes. It had been just before lunch and we were all hungry. Milt was teaching Chuck to shear and he had just graduated to his own stall and was very proud, but anxious about keeping up and doing a good job.

Randall worked as their wrangler, shoving and pushing the "woollies" from the alley into the pens beside each stall. He weighed less than 100 pounds, so I was surprised he had the strength to do what he did.

Instead of letting sheep crowd any old way into the pens, Randall selected special ones for Chuck. When Chuck opened his flap, five enormous, horned wethers with 2-year fleeces rushed

FIRST-TIME FLEECER. Randall, here at age 14, is shearing his first sheep. Cora says he did those first woollies very slowly and very carefully. Sheep-shearing is hard work and it's a good bet Randall wasn't smiling for long.

him. After picking himself up, Chuck looked into Milt's pen to see five docile ewes with 1-year fleeces, two of which had no belly wool. He let out an angry bellow.

I was tying fleeces for them both, sweeping the floor of second cuts and tags between sheep, running the hydraulic sacker and dashing to the house to fix lunch in my "spare" time, so I wasn't paying attention. I glanced up just as Randall's high-pitched maniacal giggle floated from the alley.

Chuck gnashed his teeth and drove one fist grimly into the other. He blinked furiously. "You gotta do something about that kid, Mom."

I wished we hadn't laughed, but it was impossible to keep a straight face. After that, even Milt's offer to trade stalls hadn't helped. Chuck struggled through the monsters in tight-lipped anger, and the thundercloud that had settled on his face still distorted his features as he glared at Randall across my saddle horn.

Randall's bottom lip jutted belligerently. "Oh yeah?"

"Yeah," Chuck growled. "And if you do it again, you little twerp, I'll pound you."

"Just try," Randall dared.

"Knock it off," I scolded, impatient with the endless bickering. "If you can't be nice to each other, then shut up."

Chuck snorted and nudged Grey into a trot, passing Milt and turning off onto a small grassy bench. Then he kicked him into a gallop.

"I'll show him who's a jerk," Randall muttered, windmilling his arms and legs in an effort to wake up George. "Come on, you mule, go!"

"Behave," I called as he swerved around Milt. "Don't bug Chuck."

He paid no attention, whipping George with his reins, galloping past Chuck, whooping like an Indian, then streaking away. After a startled moment, Chuck whirled after him. But the bench was small and Randall had already turned. All at once, instead of pursuing, they were challenging.

With grim faces, they galloped toward each other like medieval knights in a deadly joust. At the last moment, they both swerved and pulled their horses around in a tight circle, coming face-to-face again. They shouldered their horses against each other, ramming and shouting, their faces inches apart.

"Creep!" Chuck sneered.

"Pig!" Randall spat.

Suddenly, Chuck stood up in his stirrups and poked Randall in the chest with one finger, pushing him backward.

"That'll do." Milt's voice rang out with quiet authority.

Chuck jerked back in his saddle and pulled Grey away from George with a furious scowl. He turned his back contemptuously, just as Randall swung his reins. The narrow leather strips whistled through the air and slashed Chuck across the back of the hand.

Chuck stared at the angry red welt. "Why you..."

I was paralyzed, unable to move, cold all over, my heart slamming like a caged bird.

Milt rode his horse between them. "Enough, I said." He didn't raise his voice. "You boys need to quit this fighting. You'll end up hating each other." He spoke softly, directly at them, compelling them to meet his eyes. "The memories you make now will stay with you the rest of your life, and the ones you just made aren't very pretty."

Randall dropped his eyes with a sullen shrug. Chuck rubbed his injured hand and glared.

"I'm not asking you to apologize," Milt said softly. "You were both wrong. But I want you to shake hands and forget what happened." He nudged his horse out of the way. "Go ahead. Shake hands."

Randall's bottom lip stuck out a mile. Chuck snorted derisively.

I held my breath. Their faces were so stony, implacable. Could they put this ugly confrontation behind them and forget? Why had it happened? What had caused their usual sniping to explode into such a crisis? Why were their tempers so short? Then came the question that sends women to psychiatrists.

Was it my fault?

It was so quiet on that grassy bench that when their hands finally touched, I could hear Milt's watch ticking. Even the dogs were silent. I released the breath I was holding; my hands clutched at the saddle horn in a futile attempt to stop their shaking.

"Good," Milt said with a tone of finality. "It's forgotten." He nudged Fuzzy. "Let's get going. We have work to do."

As soon as the words were out of Milt's mouth, Randall twisted George around and shot back down the trail toward home.

I turned to follow. "Let him go," Milt said. "He won't go far."

Chuck rode up beside me and we watched Randall's retreating back. He held out his injured hand, the angry welt ridged by a thin line of blood.

"Can you believe that?" Pent-up fury still smoldered in his eyes.

"No," I said sadly. "I can't believe any of it."

"Well, I'm sorry, Mom, but I don't think it was my fault."

"You know he has a quick temper." I watched Randall pull George to a stop. "And I expect more control from you because you…"

"Yeah, sure," Chuck interrupted, "because I'm older. Well,

don't expect it anymore, 'cause I'm fresh out." He turned Grey. "I'm riding ahead. See you at Lamb Camp." He bolted away.

"I'll wait for you at the summit," Milt said behind me. "I want to take a good look with the glasses." He followed Chuck.

Randall hadn't moved. He sat on George, still as a statue, a quarter mile away. He looked small and sad and alone. And I felt helpless and inadequate and frightened.

Was being adopted the root of this? Was it jealousy over Chuck's natural birthright? I looked away from his forlorn little figure. Maybe I was making too much out of this. By evening, if they stayed true to pattern, the boys would be firm friends again and I would still be angry and upset if I tried to interfere.

I couldn't let him stay there by himself. I nosed Stormy down the trail. Randall immediately kicked George and trotted farther away.

I stopped. He obviously didn't want to talk and I wasn't going to chase him all the way home. I turned around and started for the summit. When I looked back, he was following. I breathed a sigh of relief and headed Stormy up the narrow trail without even thinking about how steep it was. At the top, Milt was standing beside Fuzzy in the rough scree, glasses pressed against his nose. I dismounted beside him.

"I counted 13," he said. "Chuck has them headed for the beach." He grabbed Fuzzy's reins. "Are you all right?" He glanced behind me. "Where's Randall?"

"He's coming." I brushed my hands across my eyes, afraid I was going to burst into tears if he so much as said one sympathetic word.

"My heart is breaking." I met his eyes and looked away.

He lifted his hand and touched my cheek. "I'm sorry."

The trail off the summit was twisting and dangerous. We had to walk the first third of the way. By the time we were able to mount, we could see Chuck far below us with the small herd.

Milt swung into the saddle. "I'll go on ahead with the dogs. You take your time." He picked his way across the rocks until the terrain smoothed into tundra again, then he galloped.

I walked beside Stormy, looking over my shoulder every few

minutes until I saw Randall and George come over the top of the hill. I waved. Randall lifted his hand but made no move to come farther.

By the time I reached the last steep winding section, Milt and Chuck had the sheep pushed far out onto the gravel beach between the cliffs and the ocean and Randall was less than 100 yards behind me.

We ate our lunch in the sand flats. My sandwich tasted like sawdust, but I made myself eat it. Milt kept up a quiet monologue. He talked about the rocks lying offshore, comparing them to those off Kashega on the Bering Sea side. I appreciated his

PRIMITIVE SCULPTURE. In this quiet land, the beauty of an offshore rock formation becomes a topic of conversation.

effort, but my mind was on Randall, who stood across the creek from us, eating his lunch behind the horses.

When we finished, we started back across the sand. Randall pulled George away.

"Wait a minute, Randall," Milt called. "I'm changing the game plan."

Randall stopped and turned reluctantly, his face still white and sullen. "What?"

"I want you and Mama to chase the sheep around the beach until you get to the first big ravine. You can't miss it. It's the only possible place they can get off the beach and it's a real steep climb." Milt caught George's halter rope. "Chuck and I will take the horses and head them off at the top." He looked over at me. "Then we'll push them against the drift fence and hope we can get them home."

I nodded, relieved that the boys would have breathing space apart and glad Milt had paired Randall with me. Maybe I could find out what was bothering him.

"Let's go," I said casually and set off across the gravel.

For the first few minutes he followed without speaking. We caught up to the sheep and started them along the beach. Then he pulled on my sleeve. When I looked around, he held out a can of Orange Crush like an olive branch.

"Want a drink?" he asked diffidently. "I been saving it."

"Thank you." I took the can. "I love Orange Crush."

"I know." His eyes ricocheted off mine. "You said it's your favorite."

His offer broke the ice, but his expression was still desolate as he clambered over the rocks after the sheep. I racked my brain for the right words, not finding any.

"I hate Chuck," Randall said suddenly, in a tight voice.

I halted in mid-stride. "Do you really?" I asked quietly.

"He hates me."

"Just because he's angry when you play silly tricks doesn't mean he hates you."

Randall blinked and dashed a hand across his eyes. It didn't help—they filled and welled over. "I didn't mean to hit him."

"I know." I watched misery wash over his features and put my hand on his shoulder. But he wrenched away.

"I was trying to hit George," he whispered.

"I know," I repeated. "Why don't you tell Chuck that?"

Randall wiped his eyes. "He won't listen."

"He might," I encouraged. "At least you can try. Now we better get after the sheep—they're almost to the ravine."

When the sheep spied the opening, they bolted for it, thinking they had escaped. It was narrow, with steep walls. A swift-

PASTURE BEDROOM. The sleeping bag on the saddle means this sheep roundup is an overnighter. The tallest mountain on the horizon is Lone Peak. This is wild horse country, and that's just what those three dark shapes in the background are.

ly moving stream plunged between them. We rested at the bottom, looking up at the high, grassy sides.

Cliff-like outcroppings merged into a smooth slope of solid rock 400 feet above our heads to the left. On the opposite side of the creek, instead of rocks, the steep walls gave way to gentle tundra-covered earth that broke onto a wide plateau.

The sheep, living up to their reputation as incorrigibles, divided into two groups. Three animals streaked up the steep incline toward the rocks, while the remaining 10 raced across the creek and stretched out single file on an invisible trail up the side of the ravine.

"I'll chase them down," Randall said. "I'm faster." He stomped on the soda can and handed it to me.

"Don't go up on the rocks," I cautioned, stowing the can in my pocket. "It's too steep. If you can't get them back by throwing rocks, just leave them."

He groaned. "I don't want to ride over here again." He darted along the stream bank.

"Be careful!" I yelled.

He waved and climbed hand over hand up the incline after

the three animals that were now far above him. I waded the creek and followed the sharp hoofprints into the tall grass, finding the trail that wound upward. It was just wide enough for one foot, single file.

The sheep I followed were halfway to the top, with me almost up to them when two broke off down the steep hillside. Screaming threats at them, I left the trail, crab-crawling and grabbing handfuls of grass to keep from falling backward off the vertical incline, trying to get under the fleeing sheep and block their escape.

The noise, more than my physical presence, deterred them. When they heard me screeching from below, they turned around. Panting and gasping for air, I crawled straight up until I reached the trail again.

I looked across the chasm and saw Randall just under the rocks. The three sheep huddled on a pinnacle above him. He lobbed rocks and shouted until he got them off the outcropping. But instead of turning around, they inched higher onto the rock face.

"Leave them, Randall!" I yelled. "Go back to the bottom and come up on this side. There's a trail."

He waved and sidled backward.

I climbed rapidly, using the tough tussocks of grass as handholds and stepping stones. The sheep toiled above me, nearing the top. I searched along the ravine edge for riders but didn't see any. I hoped they were close. After all this work, I didn't want to lose the sheep again.

A strangled shout sounded from across the ravine. I swiveled my head like an owl and saw a blurred white shape tumble off the cliff. It turned over once in midair, slammed into the vertical hillside and bounced out over the ravine. I watched in disbelief as it plummeted straight down, disappearing into a geyser of water as it crashed into the creek.

When the water settled, the sheep didn't move. I knew it was dead. "Mom!" Randall's thin, terrified scream pierced the air and echoed off the ravine walls. "I can't move."

I tore my gaze off the dead sheep and scanned the rocks. Instead of coming down like I had told him to, he had inched midway out onto the smooth slab of rock after the sheep. I took one look at his desperate, spread-eagled position 5 feet from the edge, and started clawing my way upward.

"Hang on!" I yelled in the calmest voice I could muster. "Don't look down." *Especially don't look at that dead sheep*, I thought. "I'm coming!"

"Hurry!" he shrieked.

"Hey!" Above me I heard a shout. Chuck stood silhouetted on the skyline.

"Randall needs help!" I shouted, pointing across the chasm.

I hurtled up the remaining yards, clutching, pulling, praying. I caught up with the sheep and spooked them over the top. As I staggered to my feet, I saw Milt galloping around the head of the ravine with the horses. I sprinted toward him.

By the time we reached Chuck, he had already found a way down the face by sidestepping along a break in the rock above Randall. I flung myself down and peered over the edge. Randall's ashen face stared up at Chuck.

"I can't move," he whispered, panic clogging his voice. "My boots are too slick."

Chuck wedged his feet in the crevice and slid his torso across the rock, stretching out his arm. "Grab my hand, you stupid kid."

"I can't let go!" Randall gasped. "I'll fall!"

Chuck wiggled his arm another inch closer, the vivid welt on his hand standing out in sharp relief. "No, I won't let you."

Slowly, imperceptibly, Randall's hand relaxed and began the long crawl toward Chuck's. I heard a scraping and Randall's voice cry out, "I'm slipping! I'm slipping!"

With a lightning heave, Chuck lunged another inch and closed his hand around Randall's. "I got you, little brother." He jerked him off the rock. "You're safe."

Chapter Twenty-One

IN THE BLINK OF AN EYE

Astrange sound woke me. I swam up through layers of sleep, blinking at the unfamiliar rope springs above my face. When I moved, a shaft of agony shot through my whole body and I remembered.

Rubbing my gritty, swollen eyes, the inside of Kismaliuk line cabin came into focus: tin walls, rusty potbellied stove, narrow bunks and plywood table. They all looked ghostly and insubstantial—part of a dream.

I shifted my backside off a bump in the fleece pad under me and winced. My pain was no dream. After 2 months in the saddle, I thought I was over saddle sores. But yesterday had been 35 miles of jolting, breakneck riding. Any skin I had left on my tailbone was microscopic.

I peered through the pre-dawn dimness toward the bunks nailed to the opposite wall. The sound was Milt unzipping his sleeping bag. The lump in the top bunk that was Chuck didn't move. Above me, Randall mumbled in his sleep.

"It's not light," I groaned. "Is it time to get up?"

"Just checking the horses," Milt whispered. "Go back to sleep."

He pulled on his boots and crept outside. But I was awake, and after the door closed, I crossed my arms behind my head and stared at the blank window.

Yesterday I had seen the first fireweed bloom. After we had crossed the tidal flats, where Milt stashed his bamboo pole inside the rusted hulk of a World War II barge, we were all tightening our cinches when I noticed the bright magenta column; a beautiful, quelling portent.

Where had summer gone? It seemed like only a week ago Chuck had rescued Randall from the cliff; yet it had been 6. The days and weeks had blurred, filled with sheep and cattle and coal and firewood.

Now, this first week in September, 700 sheared sheep dotted the hills and 7,200 pounds of wool waited in our warehouse to be picked up by the *North Star III* in October.

Three hundred cattle now swished flies with cropped tails, and 90 new calves were earmarked or castrated.

We'd hauled 36,000 pounds of coal across the bay in 50-pound sacks, and the yard was full of cedar, fir and cottonwood, ready to be made into kindling, firewood and smokehouse chips.

We had ridden at least 2 million miles.

I discovered I loved the fog—it meant we couldn't see. I spent those days in feverish bouts of baking and cleaning, with hours stolen to pick flowers and dye the lumpy, uneven yarn I spun between trips to the oven and clothesline.

The door creaked on leather hinges and Milt slipped back inside with an armful of driftwood. "Fog to the waterline," he sighed, using his pocketknife to whittle shavings into the stove. "Maybe it'll lift."

"Let's hope so." I felt inside my sleeping bag for my jeans and pulled them over my raw flesh. "I would hate for yesterday to be in vain."

This was our last big drive. We had already covered most of this country on day trips, going into one cove at a time, bringing everything we could find back to the corrals. This final sweep was cleanup.

Yesterday we had ridden northeast along the Bering Sea side of the island, flushing out pockets of unmarked cattle and pushing them south to the Pacific side. Today we planned to ride across the island, rounding up Kismaliuk Valley, Six Lake Basin and Buttress Point, then spend the night in Riding Cove and make the trip back to headquarters the next day, picking up the little bunches we had spooked off the Bering Sea side.

It sounded ambitious. I put my feet on the floor and fell to my knees, disguising a blood-curdling yell as a whimper. It sounded impossible!

"Whoa!" Milt hauled me upright. "You'll feel better after a cup of coffee." He handed me a tin can. "Remember where the creek is?"

Outside, the fog enveloped me in a thick, cool curtain. I found the creek by listening to the rushing water. Yesterday had been warm, with an occasional glimpse of sun through the clouds. Sometime during the night I had heard the wind howling and rain peppering the tin roof. And now this opaque blanket. We would never see cattle unless we bumped into them.

Everyone was up and dressed by the time I got back. We ate a breakfast of oatmeal cooked in hot chocolate and drank scalding cups of coffee and tea, then rolled up sleeping bags, put on rain gear and saddled our horses.

The fog was a wet sponge, dripping water on everything it touched. Big drops condensed on my hat, rolled off and splashed on my shoulders. It closed us in; tiny atoms in a planet of nothingness, as if we were the only people in the whole universe.

Close behind us, the horses' steamy breath mingled with the fog, their iron shoes hitting the rocks in muffled, ghostly music. I shivered and pulled my rain hood around my face.

PACIFIC COAST. On the Pacific Ocean side of the island, Cape Aiak, nicknamed "The Hat" for obvious reasons, is topped with fog in this view from Lamb Camp. Fog in the "crown" of The Hat only adds to what Cora calls the "otherworld" feeling along this coast.

"This should melt off when the sun comes up, but until then, stay right behind me," Milt told the boys as he boosted me into the saddle. "I don't want anyone wandering off."

"You heard him," I warned. "Get in front of me."

"Sheesh, Mom," Randall guffawed. "You're the one who'll get lost."

But he didn't argue, and guided George up behind Fuzzy.

"You go next, Mom." Chuck swung into his saddle and turned out so I could pass. "I'll keep up."

"Okay."

I nosed Stormy in behind Randall and we started off along Kismaliuk Creek. The fog was close and personal, almost a physical entity. I could barely make out Randall's shadowy form, and our horses were almost touching. I couldn't see Milt at all. To see Chuck, I had to stand up in my stirrups and look under my arm. If I turned my head, all I could see was the inside of my rain hood.

For miles along the creek we followed Milt with blind trust, our horses picking their own way through terrain we couldn't see from atop their backs. By now I was used to Stormy's frequent stumbling and jolting gait. All the horses had trouble with their footing in the uneven tundra.

I just hoped he wouldn't see anything in the fog to scare him. I'd be lost in a second if he bolted the way he usually did when he heard a sudden sound or saw something blowing in the wind, like paper or plastic from the beach.

Through the fog I thought I heard a bull bellow—high and wavering, like bagpipes. I saw Stormy's ears prick up and I tightened my grip on the saddle horn. We started climbing. The horses' hooves clinked in the scree. I saw glimpses of slick, black boulders on both sides.

And then we were out of it; riding into sunshine, blue sky and high buttermilk clouds and the summit of Six Lake Basin. Below us spread an oasis of green tranquility, strewn with small lakes and edged with the silvery blue Pacific Ocean. Cattle dotted the landscape in scattered bunches. I heard the bull again; closer this time, a hoarse trumpet challenge, then an answering bellow from across the hills.

MYSTERIOUS MARSH. The cabins in the distance are what's left of an abandoned village. One cabin is still used as a line shack, but the area remains a strange, superstitious place. Cora says the dogs won't sleep out when the family spends a night here.

We didn't speak. After working with Milt all summer, we knew what to do. He motioned Chuck and me to take the left side, then beckoned to Randall and they sauntered their horses toward the right.

We did not holler. Milt didn't like "yippee-ki-yi" cowboys. These cattle were 20 miles from headquarters. Some had never seen a man on horseback before. Those that had escaped previous roundups were wild and spooky. Any hollering or galloping would send them stampeding in all directions and right over the top of anyone who happened to be in the way.

We wanted them to run, but in the right direction, toward headquarters. So we fanned out behind them in a wide half-circle and walked our horses closer. We got within a quarter mile before they noticed us, picked up their tails and bolted. We kicked our horses into a gallop and dashed after them full-tilt until we were right on their heels.

It was perfect. They went as if we had drawn them a map—straight down the trail for Riding Cove. Stormy was an efficient cattle horse on his own. During galloping pursuits I simply hung on and hoped nothing charged me. Within 5 minutes the 40-odd animals were strung along the trail with an old black cow and her calf in the lead.

I noticed one huge Hereford bull with a smooth, curly forehead. Good, no horns. He must have been the one bellowing. There were several older calves and three tiny ones, under a month old, all bawling and following after their mothers. Several ponderous, slow moving steers brought up the rear.

Milt signaled Chuck to swing out wide to the north and get ahead of the lead. Then he and Randall left me following the trail while they scouted the small ridges and valleys for hidden bunches.

I hated this part, hated being responsible for the whole herd. I was always so afraid I would lose them all, and I some-

HEREFORD BULLS, like these two warily eyeing the camera, gave Cora and Randall quite a scare during the summer's last big roundup.

times did. This bunch seemed pretty docile. None of them turned back on me or attempted an escape. Then, from off to the left where Milt and Randall had ridden, I heard a bull bellow, hoarse and angry. Over a ridge spilled about 20 head—the bull, another hornless Hereford, snorting and bellowing in front.

With a tremendous bawl, the bull in my herd broke away and challenged the interloper. They rushed toward each other, bellowing threats, snorting, tossing their heads. When they were 20 feet apart, they stopped and pawed the ground, kicking chunks of tundra the size of saddle blankets over their shoulders.

Then they advanced upon each other, still keening like a funeral dirge. When they were close enough to touch, they circled and rushed, slamming their mammoth heads into each other's sides, ramming and twisting. Evenly matched, they pushed each other back and forth with their battering-ram heads.

The bull from my herd had the weight advantage, and when he charged, his head would lift his opponent's back legs off the ground. But the smaller one was quicker, twisting and evading with a litheness that amazed me in such a great creature. Thank God neither had horns.

Milt and Randall detoured around the melee and joined their cattle with mine. "We'll leave those bulls." Milt rode alongside. "Come on." He gave me an encouraging smile and loped away.

I nudged Stormy closer to the herd. They were slowing down now; the big steers had their tongues hanging out. I turned once and looked back at the fighting bulls. They were still at it, bellowing and snorting. I breathed a sigh of relief. Good riddance.

I saw Chuck in the distance, on the skyline like an Indian sentinel. He had just climbed out of the deep ravine between Prong Tower and Lance Point where the lead animals were now disappearing. Great! He was in place to head off anything.

I concentrated on the rear; riding up on laggards, pushing them to keep up with the main bunch. Milt stayed on one side and Randall on the other, starting to yell now and urge the animals down the steep trail, already muddy, toward the creek to the bottom. Next to the creek, the cattle churned the trail into a river of mud.

Stormy sank to his knees with every step. Below me, I saw Milt dismount and lead Fuzzy across. I tried to memorize his exact steps. Almost to the bottom, Stormy slid sideways and I jumped off into the oozing soup. Behind me, Randall angled George to safer ground.

I took one step and pulled my foot out of my boot. Holding Stormy's reins with one hand, I balanced precariously like a stork on one leg and grabbed for my disappearing boot.

"Hey, Mom!" Randall called. "What's the matter with Milt?"

I looked up, tottering and weaving, clutching Stormy for support. Halfway up the opposite bank, Milt gestured wildly, swinging his arms above his head and pointing. I glanced behind me.

Just topping the crest of the ravine were the two bulls. Their heads down like charging rhinos, they advanced on me at a gallop. I froze. In a flash, Randall was off George and pushed me out of the way. He grabbed my boot and threw himself under George, just as 4,000 pounds of uncaring destruction thundered past, veering around the horses, right where we had been seconds before. All the breath left my lungs in a whoosh. I sat down in a trembling heap.

Randall giggled. "Want your boot?" He still hunkered under George's belly, his obsidian eyes dancing with unholy merriment. "Here." He sailed it across the trail.

"Thanks," I whispered, upending it and digging out the mud. "I was too scared to move."

"No big deal," he said offhandedly, crawling between George's unresisting legs. "We better catch up." He boosted me into the saddle and led both our horses across the creek.

Milt laughed, too, when he saw me. "They weren't after you," he comforted, scraping the mud off my face. "They just didn't want to lose their harem."

We reached Riding Cove before dark and drove our cattle, about 65 head, into a cul-de-sac between the cliffs of Buttress Point and the Pacific Ocean. There was grass there, and water, and they would have to come past our camp to escape. The two bulls had dropped behind, weary from fighting and too tired to keep up.

We unsaddled the horses and picketed them by tying their stake ropes to logs we found on the beach so they could graze during the night. We used the heaviest logs we could pull between us so the horses wouldn't drag them around or get away.

There was no cabin here. Tidal waves on the Pacific side washed them away during the early days and Milt hadn't rebuilt.

THAT'S A WORD? *When the cattle are safely in, and the bulls back with their ladies, it's time to relax and warm up in the friendly ranch house kitchen. Cora ponders the Scrabble board, probably wondering if Randall learned his lessons too well.*

But our driftwood log hadn't moved. It was a huge, gnarled trunk, 3 feet high and 30 feet long, wedged into the beach gravel 50 feet from the water. With a tarp draped over it and secured with rocks, it made a dry sleeping place, with room for the boys on one side and Milt and me on the other. While we did this, Chuck started a fire and Randall went for water.

In a few minutes, Randall came running back, splashing water out of his can with every step. "Fish!" he shouted. "The stream is full." His face, flushed with excitement and still streaked with mud, showed no signs of weariness. "Come on, Chuck. Let's go catch some."

Chuck dropped his armload of sticks and they dashed away. I watched them running side by side down the beach toward the stream.

"Friends one day, foes the next," Milt said quietly beside me.

I nodded. "But they are growing up," I murmured. "They are friends more than they are foes."

Lately, Chuck had been teaching Randall to play chess.

They played after supper almost every night, their heads close together at the end of the kitchen table, sometimes silent in concentration, other times exploding in arguments and taunts.

"I wonder how they'll catch a fish?" Milt said as he gathered Chuck's tumbled sticks and laid them on the fire as the boys' voices rose and fell in the distance. Out of the corner of my eye I saw them racing up the creek with a blue plastic basket, the kind used on the floating processors and always in plentiful supply on the beach.

"They probably won't," I answered, digging out our sparse supper from my saddlebags. I put the water can among the coals and handed Milt a sandwich. We ate with our backs against the big log and our faces to the fire. When the water boiled, I made tea and we shared a cup.

But I was wrong. We hadn't finished our tea when they stalked proudly into the firelight, carrying a silver salmon between them on a ragged rope. They rigged up a tripod of wet sticks and suspended the fish over the coals.

"Boy, that was fun!" Randall exclaimed. "First I tried to catch one with my hand, but they were too fast. Then Chuck…"

"Let me tell," Chuck interrupted. "I almost had one with a rock. The fish were so thick I didn't think I could miss, but I did. They whiz by fast."

"Yeah," Randall chimed in, triumph bursting from every pore. "But we caught one with the basket and it was my idea."

"Slick, too," Chuck conceded. "I jammed it in a narrow bend and Randall chased the fish in."

A year ago I would have made an object lesson out of the teamwork they demonstrated. But not now. I had learned a few things myself over the past year.

"How long 'til it's cooked?" Randall asked, disdaining my offered sandwich.

"When it falls into the coals," Milt laughed. "Then it's done."

After 10 minutes, the boys decided they would have a sandwich. They ate in great gulps and swallowed the rest of the tea, pushing sticks into the coals and watching the flame lick at the salmon's tail. An aroma of wood smoke and sea and roasting fish wafted through the air.

With a sudden, satisfying plop, the sticks collapsed and dropped the salmon into the coals. With much shouting of advice from Randall and warnings from me about fire, Chuck used the edge of the water can to shovel the fish onto a nearby rock.

They brushed off the ashes, taking off most of the charred skin. Beneath was moist, flaky, well-done fish.

"Want some?" Chuck brought us a big chunk, juggling it from hand to hand and blowing on it. "It's hot."

We ate the grubby offering and pronounced it perfect, then left them to their feast and hiked along the beach in the twilight to check on the cattle.

They were scattered about the small alcove, cows and calves together and the big steers lying down and cropping at the tall grass.

"We're getting a lot of big steers," I commented. "They're almost too old to slaughter for beef."

"Yeah, the herd has finally outgrown the market," Milt answered. "Fifteen years ago, when I shipped in the Hereford stock to strengthen the old Russian dairy remnants that were here when I came, I never dreamed they would multiply this fast." He rubbed his jaw. "I'll talk to the crab boat skippers and see if I can't sell these old ones for bait."

"That would be a perfect solution." We sat on a smooth boulder and watched the surf roll in lazy surges onto the beach. "We wouldn't have to transport them anywhere and we could clean up the entire herd." I went on eagerly discussing how we could do it, where we could contain them and how to overcome any difficulties.

"Don't get your hopes up," Milt warned. "I know they use red meat farther out west, but around here it's still cod and herring."

Long after we had crawled into our sleeping bags under the tarp, I was still thinking about it. Beside me, Milt shifted restlessly. He never mentioned pain to me, never complained about the long hours in the saddle. How uncomfortable it must be with his steel hip joint; how tiring. I knew he was contemplating having his other hip replaced. I was so glad to see the boys take hold; do a man's work. I wished I were more help.

I dozed off, but my sleep was light. My stiff muscles, exacerbated by the rocks under me, woke me up every time I moved. When I heard the sound of galloping hooves, I was instantly alert.

Earthquake? The ground shuddered and the thunderous pounding overrode the hoofbeats. No! The cattle were escaping! I clutched Milt and shook him awake. He jerked upright and ran his head into the tarp. "Listen," I whispered.

"The horses," he muttered, his voice heavy with sleep. The pounding got closer. I heard the rocks clatter and rumble. Milt tore out of his bag. "They're dragging the logs!" he shouted. "Get up!" He rolled across me and burst out from under the tarp, anchor ropes spewing in all directions.

The high, frightened whine of a horse's whinny whistled past us. A sudden vision of the size of those logs bearing down on us exploded across my brain. My stiffness forgotten, I hurtled upward out of the sleeping bag and pulled savagely at the tarp.

"Wake up!" I shrieked. "Wake up!"

Unaware of the sharp rocks, I scrambled around the log. In the moonlight I saw Randall streaking barefoot after Milt. The pounding had diminished, and the horses were now small, blurred figures on the far hillside.

"Wake up, Chuck!" I yelled at his inert form. "We have to catch the horses."

I pulled on my boots and raced across the beach, my terror of being trampled replaced by a fear of losing the horses. If they caught their logs on a rock while they were galloping full-speed, it would break their necks. Even if all they did was run until they were exhausted, we would be afoot and lose all the cattle we had worked so hard to gather. By the time I met Milt, he was returning and tears were sliding down my cheeks.

"They're gone and I can't get up that hill fast enough on these legs." Even in the dim moonlight, I saw how obviously he limped.

"Randall took off right after you," I told him, wiping my face with the back of my hand.

"He shouldn't go after them. He'll get lost." He shifted his

CHOW TIME. *The horses are grazing on summer pasture. The bay in the foreground is Milt's horse. Stormy and Pixie are the two black horses facing away from the camera. The other black horse is Dusty, who doesn't like sheep so isn't used much.*

weight off his bad leg. "I better go look for him."

Before either of us could move, we heard the sound of galloping hooves. "Oh, no," I sobbed. "They're coming back. Where can we hide?"

I turned to flee, but Milt caught my arm. "Look," he said, disbelief and admiration mingling in his voice. "Look."

At the base of the hill, I saw George in a dead run, Randall clinging like a burr to his back. Seconds later, he slid off the sweating horse in front of us.

"Sorry," he gasped. "I couldn't catch the others."

"Good job," Milt declared. "Good job." He clapped Randall on the shoulder and grabbed the halter. "I'll saddle up and ride after them."

When Milt was gone, I noticed Randall's feet—they were bare. "How could you run in your bare feet? Didn't the rocks cut?"

"Nah," he brushed off my concern. "I didn't run till I got to the tundra. Then I found George about halfway up the hill. His rope broke, so he wasn't hauling a log."

"Well, I'm proud of you." My voice cracked and I swallowed, not wanting to blubber in front of him. He didn't notice and walked back to camp.

"It was pretty fun," he said. "I liked it."

"What was fun?" Chuck's groggy voice came from behind

265

the log. "What happened to the tarp?" He had slept right through all the excitement!

After we re-stretched the tarp, I left him and Randall to weight down the corners with rocks and walked out into the meadow to wait for Milt. He wasn't long. I watched him trot off the hill and realized it must be nearly dawn because I could see him quite clearly. He had put his saddle on Fuzzy, while George brought up the rear.

"How do you ride that mule?" he asked Randall as we all re-picketed our horses along the beach line, tying the ropes to huge, gnarled timbers this time.

"I like him," Randall said, patting the rawboned nose. "He's my friend."

After we were all settled for what remained of the night, Milt turned to me and said, "How was that for adventure?"

"Adventure!" I croaked, hearing the amusement in his voice. "I was scared to death."

He put an arm out and hugged me, sleeping bag and all. "I'm sorry."

"That's what this is all about, isn't it?" I raised up on one elbow. "Not making money. Not getting away from it all. Not becoming a hermit or any of the rest of it?"

"It's about seeing what's over the next hill," he whispered, pulling me back down beside him. "I came up here in 1948 when I was 27 years old looking for adventure. But I stayed because I loved it, even after I saw what was over every hill, knew every one of them like the back of my hand. Every day there is a newness about it; something different, something good."

He reached for my hand. "Sometimes I wonder if that is how MacIntosch felt. He was the Scotsman who started the ranch here in 1918 or thereabouts. His dream was to raise sheep in the Aleutians just like they did in the Falklands."

"It must have been a monumental task," I mused, thinking about our snug ranch house and many comforts, none of which were here then. "Yet he persevered."

"I always wished I could have met him," Milt went on. "But he and the men backing him quarreled and Mac ended up losing out." He sighed. "All that flourished from that first venture were the sheep, and even they had a tough time at first. The company, Aleutian Land and Livestock, went into receivership in the early '30s and Roy Bishop bought it. He still owned it when I came to work in '48, and it was his son who sold it to me in 1964."

"Well, he was successful," I broke in. "His family still owns Pendleton Woolen Mill in Oregon and they are known world-wide for their fabrics."

"It's family," Milt murmured, almost to himself. "Knowing it's all up to us, no outside help or interference. That makes us strong and keeps us going."

I squeezed his hand. "Don't forget the adventure."

"That, too," he whispered. "Showing you and the boys what's over the next hill is even better than seeing it for the first time."

HOME, SWEET HOME. The front gate is always unlatched and the snug ranch house is a welcome sight on a rainy day. When the chores are done, or when the weather gets its back up, the house is always there for shelter and comfort.

It took a second for the words to sink in; to realize what a beautiful compliment this quiet, private man had paid me.

"Thank you," I whispered. "I love you, too."

"Come on, you guys," Randall complained. "Go to sleep. You're making more noise than the horses did."

"Cover up your head," Milt said with a laugh. We snuggled into our own bags and finished out the night in undisturbed sleep.

The next morning, we didn't bother with a fire. We just ate sandwiches, and, being the only coffee drinker, I munched on coffee beans as we rode away from camp with our cattle.

Faint tendrils of fog clung to the hilltops and low cloud cover gave the landscape a silvery overcast and a hushed stillness so complete that I didn't want to break the spell with any human sound.

By the time we reached the head of Chernofski Creek at noon, the clouds had parted and we saw glimpses of bright sun. Even in occasional snatches, the warmth felt good on our backs as we ate lunch and congratulated each other on how well the cattle had behaved; hardly any runaways and no fights.

GOLDEN HILLS. The September sun breaks through the clouds and turns Cutter Point hill into a mountain of gold. Summers are short on Unalaska Island, and the brown grass that looks so beautiful says that fall is already here.

We were in a lighthearted mood as we pushed the herd, which had swelled to over 100 animals, along the creek toward the tidal flats. Somehow, knowing this was the last big ride lent it an extra poignancy. Like the fireweed, it signaled the end of summer.

But it was a beginning, too. We would keep some of the steers for fall butchering. In October, the *Tanya Rose* would be here with our supplies and they would want some beef. Their arrival also meant mail and Randall's school books for another year.

In the last mail delivery, Chuck got back his halibut essay with a big red "A+" on it and had decided to study for his GED through the correspondence program, so his test requirements would be coming soon.

When we reached the barge, Milt picked up his bamboo pole and we started the herd across the half mile of flat sand, now covered with a foot of water. The cattle were tired. They didn't want to go through the water.

The calves were scared, bawling and crowding around their mothers. Water splashed around our horses' legs as we urged them in among the reluctant cattle. Heads turned and tongues hung out.

We all yelled and screamed now. Milt used his pole to prod and harry. With its length, he could command a greater amount of control, and sometimes just a tap was all that was needed to start them moving.

Then, when we had everyone going, Milt cinched up on the pole and let the extra 10 feet extend behind him, under his arm. Whenever he turned his horse, anyone on either side was in danger of being knocked out of the saddle.

I looked around me at the sea of red, moving backs…at the dark water made opaque by the late-afternoon sun…at the long pole whipping crazily 2 feet from my horse…at Chuck and Randall whistling and yelling in the midst of all those cattle. It was then I wondered when I had stopped being afraid.

As we pushed the herd out of the water and through the Cutter Point gate, all I felt was pride and love and a deep sense of accomplishment. We finished the last 2 miles, fanned out be-

BOSSY AND BABY. Tulip, the milk cow, stands near her first calf. Tulip hated to be separated from her calves but that was the only way the family could get any of her rich milk for themselves. When her eighth calf was due, Tulip took off and was never seen again.

hind the lagging herd in a loose half-circle.

When we topped the last hill, the ranch headquarters came into sight below us like a tiny, sheltering outpost. I sighed a long breath of weariness and contentment, relaxing my hold on Stormy's reins and shifting my aching backside off the raw spot. At last, I really felt like cowgirl.

TWWWWANNNNGGGG!

A calf crashed through the fence 2 feet from Stormy. The sound hissed through the air like a snake. Stormy bolted straight down the hill. Caught unaware, I lurched backward and both feet came out of the stirrups. I made a frantic lunge for the saddle horn and grabbed a handful of mane instead.

Milt's astonished face streaked past, and cattle scattered in all directions. Stormy stumbled and went down on one knee. The lurch jolted me back into the saddle as he suddenly changed course and skidded along the top of a cut bank, before wheeling and charging back into the cattle from the other direction.

Randall was a blur as we whizzed up the hill almost as fast as

we had hurtled down. It couldn't have been more than 30 seconds before I unfroze enough to pull up on his head and turn
him in a tight circle.

"Whoa, Stormy. Whoa," I panted.

We shuddered to a stop beside Chuck's horse at the top of
the hill. He made an exasperated face.

"Okay, Mom. Stop showing off," he chided in a perfect imi-
tation of my sternest voice. "You're scaring the cattle."

ONCE A FAMILY, always a family. Randall and Chuck, now grown men, recently returned via floatplane to visit Milt and Cora (top). Both have chosen to remain on Unalaska Island, and make their living in town. Chuck (right) is a commercial fisherman. Randall (below) works as a longshoreman. They come to the ranch often and help with the chores.

THE WRITE STUFF. Cora's many long nights at her word processor are what led to the book you have just read.

Epilogue

THE REST OF THE STORY

Twelve years have passed since that long-ago summer. Those years brought many changes. As Chuck and Randall grew up, they made their own dreams come true.

When Randall was 14, he caught his first wild colt. By the time Chuck was 20, he had hiked the 85 miles through the Shaler Mountains to the village.

Randall is 25 years old and married to Jennifer, an artist from California. He works as a longshoreman in Unalaska village. His dreams now include a place more remote than Chernofski Sheep Ranch—interior Alaska, where there are new hills to climb and new views to watch from mountaintops.

At 29, Chuck is married to his boat. The *Prowler*, 32 feet long, wooden-hulled and weathered, is his home, his career and his life. He loves the water, and for him there will always be another cove just around the corner, another big wave, another high tide.

The passing years have brought changes for me and Milt, too. After four total hip replacements, Milt traded his horse for a track machine. And since I lost my right hand to cancer 8 years ago, I've exchanged my saddle for a typewriter.

But some things never change; like the hills around Chernofski, the Bering Sea rolling past our front door, the way we all love this place and each other. And mothers never change,

273

not really; or fathers, either. Last week, Julie Habel, a photographer for *Country* magazine, arrived to take some of the pictures you've seen in this book. Chuck and Randall, both grown men who tower over me now, flew here from Unalaska village with her.

The second day the boys were here, they saddled up two of Randall's horses for a trip across the island to the Pacific side.

As they trotted away from headquarters, Julie called, "Have fun."

"Get a good count on the Lamb Camp herd," Milt shouted after them.

"Slow down!" I shrieked.

GANG'S ALL HERE. *It's the Holmes family plus one. On the left is Julie Habel, the Iowa photographer who was sent by Country magazine to Unalaska to meet the family she had been reading about. Julie took many of the beautiful photos that appear in this book.*